D1201665

Mathematics Recovered for the Natural and Medical Sciences

Mathematics Recovered for the Natural and Medical Sciences

Dennis Rosen

Birkbeck College
University of London

CHAPMAN & HALL

London · New York · Tokyo · Melbourne · Madras

Published by Chapman & Hall, 2–6 Boundary Row, London SE1 8HN

Chapman & Hall, 2–6 Boundary Row, London SE1 8HN, UK

Chapman & Hall, 29 West 35th Street, New York NY10001, USA

Chapman & Hall Japan, Thomson Publishing Japan, Hirakawacho Nemoto Building, 7F, 1-7-11 Hirakawa-cho, Chiyoda-ku, Tokyo 102, Japan

Chapman & Hall Australia, Thomas Nelson Australia, 102 Dodds Street, South Melbourne, Victoria 3205, Australia

Chapman & Hall India, R. Seshadri, 32 Second Main Road, CIT East, Madras 600 035, India

First edition 1992

Typeset in 10/12pt Times by Interprint Limited, Malta
Printed in Great Britain by Page Brothers Ltd, Norwich

ISBN 0 412 41040 0

A catalogue record for this book is available from the British Library

Library of Congress Cataloging-in-Publication data
Rosen, Dennis, 1928–
Mathematics recovered for the natural and medical sciences
Dennis Rosen.
p. cm.
Includes index.
ISBN 0-412-41040-0 (pb.)
1. Mathematics. I. Title.
QA39.2.R6544 1991
510—dc20 91-4214
CIP

Contents

Preface

If you've lost, or think you've lost, what knowledge of and capability in mathematics you once had, why try to recover it? The answer is that you should do so if you wish to keep up with progress in the sciences into the twenty-first century: you will find it difficult to manage this without being able to follow arguments couched in mathematical terms. All the sciences – the biological, medical, economic and social ones, as well as the physical ones – are being pushed by international assault into more logical structures. Nearly all scientists are following the physicists and chemists in seeking analysis and prediction by quantified logical argument rather than by inference based on qualitative descriptions. And their medium of communication is mathematics.

Bertrand Russell set himself the task of showing that mathematics is a branch of logic. He did not completely succeed, but mathematics is certainly the language of quantified logical argument. However, the mathematics or the logic must be correctly formulated. There is a story that one of the dons at Trinity College, Cambridge, once challenged Russell at dinner: 'What's all this, Russell, about a false statement implying every other possible statement, true or false? Suppose I say that two and two makes five. Now, you prove to me that you're the Pope.' After a moment, Russell replied 'If five equals four then four equals three, three equals two, and two equals one. The Pope and I are two. Therefore the Pope and I are one.' There's a double moral in this story: you can reach all sorts of exciting conclusions by mathematical argument; but you'd better make sure that the argument is based on acceptable axioms or hypotheses and doesn't introduce errors on the way, and the proper way of making sure is to understand the mathematics.

In my experience, most students of the non-physical sciences are obliged to study mathematics (sometimes disguised as statistics) two or three times in the course of their education and promptly forget what they have learned once they have passed the requisite course examinations. But when, after graduating, they begin serious professional work, many find that, after all, their tutors had some reason for including mathematics in the syllabus and that mathematics really can be of use. Yet at that stage they may be reluctant to return to the condescending textbooks of earlier days, which in any case were sold off in the used book market at the end of the academic year.

This book has been written with such people in mind. It covers the mathematics which I have found to occur most often in typical scientific explorations, either in developing hunches in the laboratory and office, or in preparing explanations for verbal or written presentation. It is, indeed, such mathematics which many users of the book will have met and mastered before, will have forgotten, but now wish to recover. I have supposed that the users of this book haven't much time to spare from their main activity for the study of mathematics. As a result, the pace is brisk, but I hope that will prevent a loss of interest on the way.

It would be naïve on my part and on yours to suppose that methods in mathematics can be mastered, whether as discovery or recovery, without a certain amount of intellectual effort, so this book contains a variety of problems solved in detail and to be read through, or left for you to solve. I am most grateful to Derek Ray and Robert Mackintosh for the care with which they have worked through the main text and the problems and so have alerted me to many errors of typography or presentation. Any remaining errors I shall blame on my word processor.

I would like to acknowledge the following: Constable & Company for the quotation by Logan Pearsall Smith on p. 226; Faber & Faber for the quotation by Louis MacNeice on p. 117; and John Murray (Publishers) Ltd for the quotation by John Betjeman on p. 144.

Dennis Rosen
Birkbeck College
January 1991

1

Read this: a brief introduction

In the same way that biologists and medics are unable to understand the revulsion shown by most physical scientists at the sight of, or even the mention at mealtimes of, various internal bodily parts and functions – guts, subcutaneous fat, renal secretion, parturition, etc. – so physical scientists are unable to understand the terror which rises in the biologists and medics when they are faced with a differential equation, or the summation of an infinite series or an integration sign. Well, perhaps not really terror, but at least a switch-off. Whatever it is in your case, the purpose of this book is to help you conquer it and to master several mathematical skills. Nearly all of the manipulations and procedures to be dealt with in the following chapters have been presented to you before, at high school or in earlier years at university or polytechnic. Indeed you acquired them to the level of an examination pass, but then allowed yourself to forget them, or to believe you have forgotten them – else you wouldn't now be using this book.

One thing is quite certain: none of the mathematics in this book is beyond the capability of a science graduate (or undergraduate either in science or the arts, come to that). Nearly all of it was developed more than a hundred years ago and is now well within the syllabus of pre-university or first-year mathematics degree courses. It may be, though I doubt it, that quite different mental faculties are required to study the biomedical sciences on the one hand or the physico-mathematical sciences on the other. But it is quite as difficult to graduate with a first class degree in biology or medical science as to do so in physics or mathematics. So it just cannot be said in any meaningful way that physicists or mathematicians are cleverer than biologists or medics (nor vice versa) in which case it's difficult to believe that the high school and first-year degree material of one group is beyond the grasp of the other. It would be like supposing that the scientists are intrinsically unable to master the elements of a foreign language (*any* foreign language) initially studied by their confrères in the humanities faculty.

There is, however, just one important difference between being a student of mathematics and being a student of other sciences. That difference is the nature of the practicals which accompany the study. Because of the size or cost or special nature (or all three) of the apparatus and materials necessary for the most instructive experiments, the practical work of science students is for the most part confined to teaching laboratories. All students of the experimental sciences have grim memories of seemingly endless afternoons spent in those labs. In the science of mathematics, by contrast, the only necessary apparatus and material is pencil and paper, so all the mathematical experiments – seeing if problems can be solved to give answers the same as the answers at the back of the book – can be done by the student at home (or in a café or bar if home is too lonely).

Now, however tedious were those teaching lab sessions, I think that all (or very nearly all) biologists would agree that the dissections, the microscopic examinations, the biochemical analyses, the physiological experiments and so on, did indeed give cumulatively a depth of understanding of their science which would never have come with mere reading or listening to lectures. Similarly for the physicists, chemists and other experimental scientists. Hence, we can confidently dismiss the boasts of those arrogant physical scientists who claim to have mastered this or that biological science by poring over a textbook for a few days. But at the same time, the dreadful lesson for the biologist is that even the rudiments of mathematical science can be acquired only by doing the experiments: solving the problems.

However, here is a comforting thought. Learning the manipulations of elementary mathematics is in some ways like learning to ride a bicycle – once you've learned it, the skill can't really be forgotten, or at worst it lies buried somewhere, but is recoverable. This book is written to help with that recovery process as painlessly as possible. I am confident that you once, even if only briefly, had that skill, otherwise you wouldn't now be at your present academic level. As a result it is possible to offer you a book which in parts is quite severely condensed – just to remind you of what kind of situation is being examined, of what mathematical method is used, and of the mechanics of using it. There are some worked examples and several problems for you to do – enough for you to get the hang of the method, not too many to frighten you off (I hope).

I have supposed that your main aim is to reach a level of proficiency which will allow you to solve the problems which actually come your way in the course of pursuing your own science. I have therefore restricted the expositions of methods and mechanisms to the minimum, even at the risk of losing mathematical rigour or completeness. When you are ready to proceed to the next stage of advancement, you should have no difficulty in finding many appropriate books in the bookstores or libraries. The set problems, once they've progressed past the most basic matters, are restricted to run-of-the-mill situations which may be encountered in practice rather than exploring the extremes of theory and plausibility. However, most of the problems are unvarnished mathematical ones: I have made no attempt to disguise a mathematical wolf as a biological sheep.

Having written that, I must admit that the first couple of chapters may in fact seem rather wordy. The reason is that they deal with several fundamental ideas and I suspect that the reason why your progress in mathematics has been limited is that you are to some extent confused about some of the fundamentals. So I've expanded a bit about some very basic ideas, although even then I haven't so much as mentioned most of the tricky points which real mathematicians have spent their energies elucidating.

There is another matter I should point out. Although all the mathematics covered in this book appears in many places in the biological sciences, a special goal of the book is to go as far as getting a glimpse of the method and use of Fourier transforms. The reason is that, with the availability of computers, Fourier transforms have become the key to some of the most productive analytical, practical and experimental techniques to have been developed over the past two or three decades; these techniques include X-ray diffraction studies of macromolecular structure, biological signal analysis, computerized tomographic (CT) scanning, biological control theory, high-precision infra-red spectroscopy, and several others. Fourier transforms are becoming as much of a tool for the biologist as were formerly the microscope and the dissecting kit. But, once again, the way to learn to use this tool is to do the problems.

The problems are the experiments for you to do in your practical class, which is held (probably) at your desk or table at home. These problem–experiments enable you to repeat other people's experience, in the same way as you did in dissecting a dogfish or staining a transverse section of *Zea* to show the cambium. No penalties are imposed if you don't do the problems, but if you don't you almost certainly will not do any better with mathematics this time than you did last time.

That last sentence may cause you an immediate yawn. But consider this. Academics older than you, and those who write books, are rarely cleverer than you, they are merely older. However, being older, they have more experience than you have. Experience, of course, is neither here nor there if it's irrelevant. Yet in the present case, it is the uniform experience of generations of mathematics teachers that if students of mathematics don't do the problems they get nowhere with the subject.

You are certainly adult enough to choose your own path and make your own choice, but if your choice has been for a path leading to recovery of your mathematical capabilities, don't deceive yourself by imagining that you can reach your goal without doing the exercises (and, it goes without saying, doing them correctly). Of course, you may not like this book and my guide for you. If not, by all means (a) find a better book and (b) write to let me know what's wrong with this one. But do not give up your aim. Modern biology is becoming steadily more mathematicized; if you want to understand modern biology you will almost certainly need mathematical assistance. One way or another, you must recover your mathematical skills.

2

Numbers, signs and symbols

Don't be frightened by the way mathematicians use symbols. Remember that you are pretty adept in the use of a large variety of symbols – more about that in a moment – and that symbols are used in mathematics as tools for cutting quickly through an argument to reach the solution to a problem. So learn to ride the tiger of mathematical symbolism in order to improve your efficiency in getting to answers.

Consider first some symbols found daily around you. The Union Jack, the Stars and Stripes, the Tricolour and other flags can stand for countries, peoples or histories. Schools, universities and polytechnics often have symbols – typically a shield with armorial bearings – which represent and announce them at the top of their letter paper or inside the covers of books in the library. Large companies have similar devices which they call logos and sports clubs and others call them badges. A sign outside a shop may convey that the place is a pawnbroker's or a barber's or that a certain type of credit card is accepted there. This list could go on and on and hardly any of it causes you distress or puzzlement: you take the meanings and usages in your stride. You can do the same with the signs and symbols of mathematics.

Now consider 8, which stands for 'eight', which may be standing for :::: or oranges or trees or whatever. It will not worry you if 8 is replaced by VIII; nor that the same thing is written as ٨ in Arabic and as ח in ancient Hebrew. Whether or not you know the symbols, you understand the symbolism. There is similarly no problem about $+$, $-$, $\times$, or $\div$ for plus, minus, times, and divide by. But pause here. In mathematical statements, multiplication is *commonly* indicated in three ways: by $\times$, by $\cdot$, or by nothing at all. If a stands for a number and b also stands for a number, then $a \times b$ means the same as $a \cdot b$, and that means the same as ab. Once again, these are symbols and usages which you have met before and which cause no problems. Similarly, division might be indicated by $a \div b$ or by a/b, or even by other, less frequently used, symbols.

Another well-known notation is that for expressing powers; the process called exponentiation. This represents $a \times a$ by a^2 (read as 'a squared'), $a \times a \times a$ by a^3 (read as 'a cubed'), and so on. *If* k *is a positive integer*, a^k (read as 'a to the k') means k instances of the number a multiplied together; and $a^0 = 1$ for any value of a except zero. If k is *not* a positive integer, the matter is more complicated and we shall meet it in the next chapter.

2.1 SOME REMINDERS ABOUT ARITHMETIC

Although the notations for the four simple arithmetical operations are well known, they need to be supplemented by brackets so that ambiguity is avoided. Allowing ambiguity into an argument is one of the surest ways of being led to a wrong answer to a problem, so kill it off at the beginning. If $a, b, c, d, \ldots$ stand for numbers, none of them zero, then a/bc might mean $(a/b) \cdot c$ or might mean $a/(bc)$. Similarly, $a + b/c + d$ means $a + (b/c) + d$ whereas $(a+b)/(c+d)$ may have been intended. The use of brackets obviously abolishes ambiguity and improves clarity; and experience with simple computer programming has alerted many people to the need for bracketing. The reservation that none of the numbers is zero was to avoid seeming to divide by zero. In division, a dividend is divided by a divisor to yield a quotient. However, the operation is *not defined* when the divisor is zero; '5 divided by zero' is a statement without meaning.

It does not matter in which order numbers are added or multiplied; that is, $a + b = b + a$ and $ab = ba$. This is called the commutative law, or property, of numbers. Similar features, which we all know from experience and hardly think about, are the associative law, which says that $a + (b + c) = (a + b) + c$ and $a \cdot (bc) = (ab) \cdot c$, and the distributive law, which says that $a \cdot (b + c) = ab + ac$. But although $ab = ba$, it is customary when one of the numbers has a specified value to put it first, i.e. to write $5x$ rather than $x5$ and similarly to write $2\pi n$ rather than $n\pi 2$ or $\pi n 2$.

2.2 NUMBER BASES

Although, when a and b are numbers, $ab = ba$, it is obviously not true that 73 equals 37. The obviousness of $73 \neq 37$ (the symbol $\neq$ means 'does not equal') comes about because of the convention we use when numbers are written with the numerals 1, 2, 3,... The convention, of course, is that the positions of the numerals from right to left in the written number are associated with increasing powers of 10. So, for example, 5678 means $5 \times 10^3 + 6 \times 10^2 + 7 \times 10^1 + 8 \times 10^0$, i.e. 5 times 1000 plus 6 times 100 plus 7 times 10 plus 8 times 1. Because of the use of 10 in this convention (coming from our possession of ten fingers) the convention is called the decimal notation. Conversely, the number 10 is called the **base** of the decimal notation for writing numbers.

Despite the convenience of using fingers for counting, there is no necessity to restrict the notation for writing numbers to the system with base 10. Division of the day into 24 hours, of hours into 60 minutes, and of circles into 360 degrees seem to be the historic relics of earlier, probably Babylonian, civilizations working with numbers to base 6 or 60; and the three-toed sloth possibly calculates with base 6. There was formerly a torture devised for children beginning to learn mathematics, to convert the writing of numbers from one bizarre base to another (e.g. convert 953 (base 23) to base 17). Such exercises are pointless except as punishment. However, for technical reasons it is sometimes convenient in working with computers to express numbers in binary (base 2), octal (base 8), or hexadecimal (base 16) notation.

1. For base 10, the numerals used are: 0, 1, 2, 3, 4, 5, 6, 7, 8, 9
2. For base 2, the numerals used are: 0, 1
3. For base 8, the numerals used are: 0, 1, 2, 3, 4, 5, 6, 7
4. For base 16, the numerals used are: 0, 1, 2, 3, 4, 5, 6, 7, 8, 9, a, b, c, d, e, f where a stands for 10, b for 11,... f for 15 when written with base 10.

Thus 1234 (base 8) means $1\times 8^3+2\times 8^2+3\times 8^1+4\times 8^0=668$ (base 10) and f00d (base 16) means $15\times 16^3+0\times 16^2+0\times 16^1+13\times 16^0=61453$ (base 10).

Conversion of numbers to and from base 10

Example 2(a) Convert hexadecimal 1cec0de to decimal notation.

Answer: The method has been indicated in the previous section, with the understanding that c (base 16) = 12 (base 10), etc.

$$\begin{aligned}\text{1cec0de (base 16)} &= 1\times 16^6+12\times 16^5+14\times 16^4+12 \\ &\quad \times 16^3+0\times 16^2+13\times 16+14 \\ &= 30327006 \quad \text{(by desk calculator).}\end{aligned}$$

Example 2(b) Express 137 (base 10) in binary notation.

Answer: Make a continued division sum of it.

$$
\begin{array}{rcrl}
 & \text{remainder} & & \\
2\underline{)137} & & & \\
2\underline{)\ 68} & 1 & \text{i.e. } 137 & = 68\times 2+1\times 2^0 \\
2\underline{)\ 34} & 0 & & = 34\times 2^2+1\times 2^0 \\
2\underline{)\ 17} & 0 & & = 17\times 2^3+1\times 2^0 \\
2\underline{)\ \ 8} & 1 & & = \ 8\times 2^4+1\times 2^3+1\times 2^0 \\
2\underline{)\ \ 4} & 0 & & = \ 4\times 2^5+1\times 2^3+1\times 2^0 \\
2\underline{)\ \ 2} & 0 & & = \ 2\times 2^6+1\times 2^3+1\times 2^0 \\
2\underline{)\ \ 1} & 0 & & = \ 1\times 2^7+1\times 2^3+1\times 2^0 \\
0 & 1 & &
\end{array}
$$

The required expression is obtained by reading the remainder column upwards, viz. 10001001 and the meaning in base 2 of this number is given by the last line of the right-hand column.

EXERCISES 2A

1. Convert to decimal notation the numbers:
   (a) 1010101 (base 2);
   (b) 54321 (base 8);
   (c) abcde (base 16).
2. Write the decimal number 97 in binary form.
3. Write the decimal number 297 in octal form.
4. Write the decimal number 72783 in hexadecimal form.

2.3 TYPES OF NUMBER

The types of number ordinarily met with are called real numbers. The name distinguishes them from complex numbers, which appear in later chapters, and transfinite numbers, which don't appear in this book. Real numbers can be positive or negative and if a is a real number then $-a$ is also a real number, so that the sum of the two is zero.

Among the numbers there are two special ones, zero and unity, such that, whatever the value of a,

$$\left.\begin{array}{r} a+0=a, \quad a\cdot 0=0 \\ a\cdot 1=a/1=a \end{array}\right\}. \tag{2.1}$$

The real numbers are classified as integers, rationals and irrationals. The integers are sometimes separated into the positive and negative integers. The rational numbers are those which can be expressed as a division, like 3/4 or $-9745/86423$ or generally a/b where a and b are positive or negative integers. Irrational numbers are ones which cannot be expressed as a simple division involving only integers: examples are the square root of 2, π, and the cube root of 5.

2.4 FUNDAMENTAL ARITHMETIC OPERATIONS

Addition and subtraction

These are so familiar that they rarely cause difficulty of understanding, although there may be difficulties in respect of the mechanics of the tasks.

Multiplication

The product of two positive numbers is positive, so that if a, b, and c are positive non-zero numbers, $a(b+c)=ab+ac$, which we met before as the distributive law in Section 2.1. The product of a positive and a negative number is negative as can be seen from the argument:

$$a \cdot (b-b) = a \cdot 0 = 0 \qquad \text{from equation (2.1)}$$

but also

$$\begin{aligned} a \cdot (b-b) &= a \cdot b + a \cdot (-b) \quad \text{by the distributive law} \\ &= 0, \qquad \text{provided that } a \cdot (-b) = -(a \cdot b). \end{aligned}$$

The purpose of my providing that argument is to give you confidence in the corresponding argument to demonstrate that the product of two negative numbers is positive, a concept which worries many people. For consider that

$$(-a)(b-b) = (-a) \cdot 0 = 0$$

and also

$$\begin{aligned} (-a)(b-b) &= (-a) \cdot b + (-a)(-b) \\ &= -ab + (-a) \cdot (-b) \\ &= 0, \qquad \text{provided that } (-a) \cdot (-b) = ab. \end{aligned}$$

Division

Division is an appreciably more complicated operation than multiplication and for nearly all daily purposes is best left to a calculator or computer which, indeed, is often the case for addition, subtraction and multiplication when the numbers are large or awkward. Division really needs a formal definition (as, strictly, do the other arithmetic operations) which I shall not give. Suffice it to say that if a dividend a is divided by a divisor b to yield a quotient c, i.e. if $a \div b = c$ or (alternative notation) $a/b = c$, then c is *defined* by $c \cdot b = a$. From this it follows that if dividend and divisor are both positive or both negative then the quotient is positive; while if one of the dividend and divisor is positive and the other is negative, then the quotient is negative. This definition also gives a hint about why division of a number a by zero is normally not allowed (i.e. has no meaning and is not defined). Since the result of the division, the quotient c, is defined by $c \cdot b = a$, it follows from the definition of the properties of zero (equation (2.1)) that if b is zero then a must be zero, whatever the value of c, so c is not specified by its definition. (Only the special case of zero divided by zero *sometimes* can be dealt with, but the value of the quotient is not then obtained by any simple means and must be determined by another route.) There is no will-o'-the-wisp to be followed here; just don't try to divide by zero.

Square roots

Because of the rules of multiplication of positive and negative numbers, if r is a square root of a positive number a, so that $r^2 = a$, then $-r$ is also a square root of a, since $(-r)^2 = r^2 = a$. We write $r = \sqrt{a}$, where the sign $\sqrt{}$ means 'positive square root'; then $-r = -\sqrt{a}$. It is sometimes convenient to indicate the two square roots by the symbolism $\pm\sqrt{a}$. No real number is the square root of a negative number since no real number, whether it is positive or negative, will, if multiplied by itself, give a product which is negative.

2.5 NUMBER PAIRS

If one is given two numbers, a and j, say, and asked to obtain the numbers s and p which are their sum and product, the procedure is straightforward: $s = a + j$ and $p = aj$. But what should one do if given two pairs of numbers, (a, b) and (j, k), and asked to obtain the pair of numbers, (s, t) which is their sum and the pair (p, q) which is their product? Answer: one can't do anything until the rules and procedures of addition and multiplication of the number pairs have been stated and defined.

Number pairs, or triplets or still larger sets, appear in various places in mathematics and they will be found in later chapters as points on a graph, as complex numbers, as vectors and as matrices. As a first exploration of the idea, look at this scheme:

definition of addition: $(a, b) + (j, k) = (ak + bj, bk)$ (2.2)

definition of multiplication: $(a, b) \cdot (j, k) = (aj, bk)$ (2.3)

system constraint: $(ap, bp) = (a, b)$ for any p except zero.

What should be the formula for division to accompany these operations of addition and multiplication? Let the quotient of (a, b) and (j, k) be (x, y) so that $(a, b)/(j, k)=(x, y)$. We take the quotient to be defined by the same procedure in principle as for single numbers (cf. Section 2.4). That is, the quotient (x, y) is defined by $(x, y)\cdot(j, k)=(a, b)$. But $(x, y)\cdot(j, k)=(xj, yk)$ so that $(xj, yk)\equiv(a, b)$. (The symbol $\equiv$ means 'identical with'; it is slightly different from $=$, as discussed at the beginning of Chapter 7.) Hence we can identify xj with a and put $x=a/j$; and similarly identify yk with b and put $y=b/k$. Thus we end up with:

$$(a, b)/(j, k)=(a/j, b/k) \tag{2.4}$$

a result which additionally requires that neither j nor k is zero.

What I have done here is to provide a notation for simple fractions which differs from the usual one. Look at

$$\frac{a}{b}+\frac{j}{k}=\frac{ak+bj}{bk}$$

$$\frac{a}{b}\cdot\frac{j}{k}=\frac{aj}{bk}$$

$$\frac{a}{b}\div\frac{j}{k}=\frac{a/j}{b/k}.$$

It can easily be seen that in the number pairs operations defined by equations (2.2) and (2.3), the first number of the pair is the numerator of the fraction and the second is the denominator. And if actual numerals are substituted for a, b, j, etc. the answers obtained in manipulating pairs in accordance with the definitions of equations (2.2) and (2.3) would be identical with the ones obtained by the usual manipulations of cross-multiplying fractions, etc.

There may seem to be no advantage in representing fractions by number pairs rather than by the usual numerator–denominator notation, but the number pair scheme emphasizes that ordinary single numbers are merely the first members of a hierarchy or family of mathematical objects. Other members – vectors, matrices and tensors, for example – can also usually be added, subtracted, multiplied, and otherwise be operated upon, although for each such object or group of objects the details of the arithmetical operations will need to be defined.

EXERCISES 2B

1. Given the rules of arithmetic operation of number pairs of equations (2.2)–(2.4), show that the pair (1, 1) is the unity term of the system and

that the pair $(0, k)$, where k is any number other than zero, is the zero term of the system, where unity and zero are as defined by equation (2.1).

2. Given the rules of addition and multiplication of a system of number pairs as:

$$\text{addition:} \quad (a, b)+(j, k)=(a+j, b+k)$$
$$\text{multiplication:} \quad (a, b)\cdot(j, k)=(aj-bk, ak+bj),$$

calculate
(a) $(5, 4)+(7, 3)$;
(b) $(0, 3)\cdot(0, 3)$;
and also
(c) show that $(1, 0)$ is the unity term of the system.
(This is the system of complex numbers discussed in Chapter 5.)

3

Basic algebra

3.1 NOTATION AND CONVENTIONS

There are several examples of notation which I have used on earlier pages and have supposed are familiar to you. An example is the use of letters – a, b, x, etc. – to stand for numbers. Because of the convention that ab indicates the product of a and b, there is a convention *not* to use two-letter symbols for numbers. (This is in contrast to computer programming usage where two or more letters for each variable address is a common practice and is even encouraged.) It is usual not to use the letter o to stand for a number because it is likely to be confused with the zero numeral 0. On the other hand, as a memory aid, it is often convenient to represent a number which is a measure of a real thing by the letter which begins the name of that thing, so m might be used for mass, h for height, r for radius, etc.

In this way, people using mathematics soon run out of letters – not just mathematicians but most scientists and economists – and in my experience it is the physical chemists who run out of symbols soonest. After the upper and lower case letters have been used up, the resorts are subscripts (e.g. r_1, r_2, r_3 for the radii of a series of circles) or prime accents (e.g. x', x''), then Greek letters and bold-face fonts (e.g. $\mathbf{a}, \mathbf{b}, \mathbf{c} \ldots$) and then, in rising desperation, gothic script, cursive scripts, and even Russian or Hebrew letters.

A quantity which takes any value within some range (or domain) of values is called a variable. (In the special case of statistics, where the quantity may have been a sample in a series of measurements, it is called a variate, as discussed in Chapter 13.) When the value of one variable, which might be called x, allows another variable, which might be called y, to be calculated, it is said that y is a function of x and also that x is the independent variable and y the dependent variable. It may be, for example, that

$$y = 2x^3 + 3x^2 - 8x + 5. \qquad (3.1)$$

If the value of the independent variable x is known, it is very easy with a calculator to work out the value of the dependent variable y. However, if only the value of y is given, it may be difficult to work out the value of x which gives rise to that value of y and indeed there may be more than one value of x resulting in the same value of y. (Try putting $x=+2$, $x=-1.5$ and $x=-2$ into equation (3.1).)

When y is a function of x, but when the details of the function have not yet been specified, it is usual to write $y=f(x)$ (read as 'y equals f of x') or $y=g(x)$ or $y=h(x)$, etc. where f (or g or h) stands for the function and x is called the argument of the function. In this notation, $f(x)$ might be the right-hand side (r.h.s.) of equation (3.1) or perhaps the r.h.s. of $y=\sin x$, or of $y=\log x+\sqrt{x}$, or whatever. Note that a function is an operation. In equation (3.1) it is the operation 'twice the cube plus three times the square minus eight times, plus five'; we have a function of x when the operation is performed on x. We say that the function maps from the domain of the independent variable into the range or domain of the dependent variable. Clearly, the operation might be performed on some other variable, for example, on p to give q if $q=f(p)$.

3.2 INDICES

As you well know, in the notation of exponentiation x^2 means $x \cdot x$, x^3 means $x \cdot x \cdot x$, and so on. The superscript is called the power or exponent of x. If a is an integer, x^a means a values of x multiplied together; it is referred to as 'x to the a' or 'x raised to the power a'. Since $(x \cdot x) \cdot (x \cdot x \cdot x)=x^5=x^{2+3}$, there is an obvious generalization that, when a and b are both positive integers,

$$x^a \cdot x^b = x^{a+b}. \tag{3.2}$$

This is clearly intelligible if a and b are positive integers, but what if they are not? It can be shown, although it is unnecessary to do so here, that equation (3.2) is a meaningful statement, whatever the values of a and b. It is, however, necessary to understand the meaning of x^a when a is not a positive integer. This can be done adequately by demonstration, but note that demonstration is evidence, not proof.

Using the appearance of equation (3.2) and putting $a=b=\frac{1}{2}$, we get $x^{\frac{1}{2}} \cdot x^{\frac{1}{2}}=x^1$. But x^1 is just x. So $x^{\frac{1}{2}}$ is that number which, multiplied by itself, results in x. In other words, $x^{\frac{1}{2}}$ means the square root of x. In the same way, $x^{\frac{1}{3}}$ means the cube root of x (since $x^{\frac{1}{3}} \cdot x^{\frac{1}{3}} \cdot x^{\frac{1}{3}}=x$). And, by extension, $x^{\frac{23}{5}}$ means the fifth root of x^{23}, etc. In general, if m and n are positive integers, $x^{m/n}$ means the nth root of x^m.

Since $x^a \cdot x^b=x^{a+b}$, it is evident that $x^{a+b}/x^b=x^a$. But also, if it is true that equation (3.2) holds for negative integers, then $x^{a+b} \cdot x^{-b}=x^a$. From this argument, we can identify x^{-q} with $1/x^q$, because dividing x^p by x^q is identical to multiplying it by $1/x^q$. It then follows that if we make $q=p$, so that the expression becomes $x^p \cdot x^{-p} = x^0$, a number x^p has been divided by itself and the result of the division, x^0, must be unity. That is, any number (other than zero) raised to the power zero has the value 1. (0^0 doesn't mean anything.)

Thus x^r has a clear meaning if r is positive or negative, integer or rational. But what if r is irrational? It can be shown that for r irrational x^r is meaningful, although the meaning can be extracted only by the use of logarithms as will be seen in the next section.

Finally in this catalogue, note that $(x^r)^s = x^{rs}$. The correctness of this statement can be appreciated by considering the example of $(x^2)^3$, which means $x^2 \cdot x^2 \cdot x^2$ and which, in turn, by what has gone before, can be written as x^6.

Example 3(a) Evaluate $(27/125)^{\frac{2}{3}}$.

Answer: Observe that $27 = 3^3$ and $125 = 5^3$. But $(3^3)^{\frac{2}{3}} = 3^2$ and $(5^3)^{\frac{2}{3}} = 5^2$. Hence the value of the expression is $3^2/5^2 = 9/25 = 0.36$.

Example 3(b) Simplify $(3x+1)^{\frac{5}{2}} - (3x+1)^{\frac{1}{2}}$.

Answer: The factor $(3x+1)^{\frac{1}{2}}$ is common to both terms, so take it out, giving $(3x+1)^{\frac{1}{2}} \cdot ((3x+1)^{\frac{4}{2}} - 1)$. But $(3x+1)^{\frac{4}{2}} = (3x+1)^2 = 9x^2 + 6x + 1$. Hence the expression is

$$(3x+1)^{\frac{1}{2}}(9x^2 + 6x + 1 - 1) = (3x+1)^{\frac{1}{2}}(9x^2 + 6x) = 3x(3x+2)(3x+1)^{\frac{1}{2}}.$$

EXERCISES 3A

1. Evaluate
   (a) $(81/256)^{\frac{3}{4}}$;
   (b) $((\sqrt{3})/4)^{\frac{3}{2}}$;
   (c) $(178/356)^{-3}$.
2. Simplify $3(a+2b)^3 - 2(a+2b)^2 - (a+2b)$.

3.3 LOGARITHMS

The logarithm is a particularly important mathematical function, the inverse of exponentiation. A logarithm of a number is taken with respect to a base. If the result of taking the logarithm of y with respect to base b is x, this is written as

$$\log_b y = x$$

and its interpretation is

$$y = b^x;$$

that is, the logarithm of a number to a stated base is the power to which the base must be raised to equal the number. The original number, y, is called the antilogarithm of the power, x.

If there are two numbers $y_1 = b^{x_1}$ and $y_2 = b^{x_2}$, then

$$y_1 \cdot y_2 = b^{x_1} \cdot b^{x_2} = b^{x_1 + x_2}$$

so that

$$\log_b(y_1 \cdot y_2) = x_1 + x_2 = \log_b y_1 + \log_b y_2.$$

Thus the logarithm of the product of two numbers equals the sum of the logarithms of each. By a corresponding argument,

$$\log_b(y_1/y_2) = \log_b y_1 - \log_b y_2.$$

These arguments are readily extended to products and divisions involving more than two numbers.

It also now follows, since x^2 means $x \cdot x$, that $\log_b x^2 = 2 \log_b x$ and, in general, $\log_b x^r = r \cdot \log_b x$. This last equation holds for any value of r – positive or negative, rational or irrational – and so gives meaning to the quantity x^r when r is irrational (cf. Section 3.2).

For about 300 years scientists, economists, mathematicians, and others have used the method of logarithms to multiply or divide large and awkward numbers. To do so, one looks up the logarithms of the numbers in 'log tables', adds or subtracts and then extracts the answer from antilog tables. People required to do sums of this kind were attached to the method because addition and subtraction are so much easier than multiplication and division. However, it is much easier still to use an electronic calculator which is now the method of choice for efficient calculation by everyone. Nevertheless, logarithms as functions appear in many places in biological and other sciences and their mathematical features must still be understood.

One feature it is particularly useful to know is how to convert the logarithm of a number from one base to another. If $\log_c y = x$, then $y = c^x$. Suppose that $\log_b c = D$ so that $c = b^D$. Then we have that

$$\begin{aligned} y &= c^x = (b^D)^x \\ &= b^{Dx}. \end{aligned}$$

So by definition,

$$\begin{aligned} \log_b y &= D \cdot x \\ &= \log_b c \cdot \log_c y. \end{aligned} \tag{3.3}$$

In normal scientific experience, only two numbers are in use as the base of logarithms: 'common logarithms' are calculated to base 10 and 'natural logarithms' are calculated to base e, where e, called the exponential number (or just e) has the value 2.71828 ... and is an irrational number with a status similar to the status of π. Logarithms to base 10 are nearly always (except in mathematicians' books) written just as log, with no base indicated; and logarithms to base e are often written as ln, but occasionally as $\log_e$. It is usual to write $\log_{10}e$ as M, i.e. $e = 10^M$. Then, from equation (3.3), to convert from natural to common logarithms we have

$$\log_{10} y = \log_{10} e \cdot \log_e y = M \log_e y$$

or, as commonly written,

$$\log y = M \ln y.$$

Conversely,

$$\ln y = (1/M) \log y.$$

Reference books show that $M = 0.43429\ldots$ and $1/M = 2.30258\ldots$

It should be noted that the logarithmic function is defined only for an argument greater than zero. There is no meaning to $\log_b x$ if x is zero or negative, whatever the value of the base.

3.4 FACTORIALS

The product of the first n integers is called factorial n and is written $n!$ or $\underline{n|}$; the former of these two notations is now the more common because it is easier for printers and typists to display. Thus $n! = n \cdot (n-1) \cdot (n-2) \ldots 4 \cdot 3 \cdot 2 \cdot 1$. The factorial function is another function with meaning only for a restricted range of argument, namely that the argument must be a positive integer; it is meaningless for non-integer or negative arguments. However, note that, for reasons unnecessary to pursue here, factorial zero, i.e. 0!, is *defined* as having a value of 1.

3.5 PERMUTATIONS AND COMBINATIONS

If there are a number of objects, for example, knife, fork, spoon, cup, saucer, plate, dish, they can be listed or physically placed in an order. But that order can be changed, or permuted, into another order. A permutation is one among the many orderings.

The total number of possible permutations of a group of objects can be readily calculated. Suppose there are n objects to be put in order, and imagine n boxes laid out in a row into which to put the objects. There are then n choices of which object to put in the first box. Once that choice has been made, there are $n-1$ objects left from which to choose the one to put in the second box; and that remains true whichever object was the first one chosen. So the first two boxes can be filled in $n\cdot(n-1)$ ways. There are now $n-2$ objects from which to choose the one to put into the third box, so that the first three boxes can be filled in $n(n-1)(n-2)$ ways. This argument continues until there are only two objects left from which to choose one for the last-but-one box and then only one object for the last box. Thus the total number of ways in which the boxes can be filled, that is, the total number of permutations of n objects, is factorial n, because $n(n-1)(n-2)\ldots 3\cdot 2\cdot 1 \equiv n!$

Example 3(c) In how many ways can the following seven coins be arranged in a line: £1, 50p, 20p, 10p, 5p, 2p, 1p?

Answer: 7! i.e. 5040 ways.

Example 3(d) If another penny, indistinguishable from the first, is added to the coins in example 3(c), in how many distinguishable ways can the coins now be arranged?

Answer: If the coins were all distinguishable, there would be $8! = 40320$ permutations. But these orderings can be taken in pairs with the two pennies interchanging positions; and since these pairs will look identical, the number of distinguishable permutations is $8!/2 = 20160$.

Example 3(e) How many ways are there of permuting the order of 8 guests, 4 men and 4 women, round a circular dining table?

Answer: Set them out in a line; then there will be 8! permutations. Now group these permutations into octets, since at the circular table $abcdefgh = bcdefgha = cdefghab = \ldots$. Hence the required number of permutations is $8!/8 = 7!$

Example 3(f) Repeat the last calculation with the proviso that men and women are seated alternately.

Answer: Again set them out in a line – man, woman, man, woman, ... There are 4! ways of arranging the men's places and 4! of arranging the women's. But since the arrangement could have been woman, man, woman, man, ... there are $2(4!)^2$ linear arrangements. Now, as in example 3(e), take octets to allow for the circular table, giving a total of $2(4!)^2/8 = 4!\cdot 3!$ arrangements.

Rather than permuting all n objects, the task could have been limited to choosing and permuting just r of them. From the account in the previous paragraph, the number of ways of arranging n things three at a time is $n(n-1)(n-2)$. If the argument is continued in detail, it is found that what is called the number of permutations of n objects r at a time, expressed by the notation ^nP_r, is $n(n-1)(n-2)\ldots(n-(r-1))$. That is,

$$^nP_r=n(n-1)(n-2)\ldots(n-r+1)$$
$$=\frac{n(n-1)(n-2)\ldots(n-r+1)(n-r)\ldots 3\cdot 2\cdot 1}{(n-r)(n-r-1)\ldots 3\cdot 2\cdot 1}.$$

Hence the general formula is that $^nP_r=n!/(n-r)!$. Since 0! has been defined to have the value 1, it follows that $^nP_n=n!$, which we found earlier.

Returning to the game of filling boxes with table objects, the number of ways of filling, say, the first three boxes is $^nP_3$. If we now say that the order in which the boxes are filled doesn't matter, we have the task of calculating what is called the number of combinations from n things, three at a time, expressed by the notation $^nC_3$. The situation here is that (knife, spoon, cup) counts the same as (cup, spoon, knife) and as (cup, knife, spoon), etc., but these are different from the group (knife, spoon, plate). Now, we know that there are 3!, i.e. 6, ways of arranging (or permuting) three objects. Hence, the $^nP_3$ permutations must in this case be grouped in sextets and $^nC_3$ is the number of such sextets, so it has the value $^nP_3/3!$.

The symbol for the general case of the number of combinations of r things at a time out of a total of n things is ^nC_r. The argument just given for three things extends to r things so that

$$^nC_r=\frac{^nP_r}{r!}=\frac{n!}{r!(n-r)!}. \tag{3.4}$$

Note that

$$^nC_1=\frac{n!}{1!(n-1)!}=\frac{n}{1}$$

$$^nC_2=\frac{n!}{2!(n-2)!}=\frac{n(n-1)}{1\cdot 2}$$

$$^nC_3=\frac{n!}{3!(n-3)!}=\frac{n(n-1)(n-2)}{1\cdot 2\cdot 3},$$

and so on. In these formulae, each of the fractions in the last column has the same number of elements in the numerator as in the denominator; ensuring that this is the case provides a useful guard against errors when such quantities need to be written out. Note also that $^nC_{n-r}={}^nC_r$.

That is effectively all there is to permutations and combinations, but problems can be very difficult indeed.

Example 3(g) In how many ways may you be dealt a hand at bridge?

Answer: ${}^{52}C_{13} = 6.35 \times 10^{11}$.

Example 3(h) In how many ways might cards be dealt at a bridge table between the four players N, E, S, and W?

Answer: First find the number of distributions of hands between four places – say A, B, C, and D – and then permute these distributions between N, E, S, and W. Give any 13 cards to A, then any 13 of the remaining 39 to B, then any 13 of the remaining 26 to C and what's left to D. Thus there are ${}^{52}C_{13} \cdot {}^{39}C_{13} \cdot {}^{26}C_{13} \cdot {}^{13}C_{13}$ distributions of hands, i.e. about 5.4×10^{28} possibilities to be permuted between N, E, S, and W (4! permutations for each distribution) giving a total of approx. 1.3×10^{30} deals.

EXERCISES 3B

1. How many ways are there of arranging the table in example 3(f) if one of the men and one of the women are on no account to be put next to each other?
2. A sales person must visit six towns, A, B, C, D, E, and F. It is convenient either to start the tour at A and finish at F or else to start at F and finish at A, but the order of the others is immaterial. How many routes might be followed?
3. A tennis club has 15 women members and 9 men and a committee of 5 is to be chosen. In how many ways might this be done? In how many ways might this be done if the committee is to contain
   (a) at least one man?
   (b) at least one man and at least one woman?
   (c) not more than two men?

3.6 MATHEMATICAL INDUCTION

Experimental scientists search for evidence to support a theory, but mathematicians search for proof to validate a statement. Proof is an elusive quality and there are strict rules for ensuring it. One set of rules is the methodology called mathematical induction, which works like this:

1. Suppose there is a regular sequence of statements, numbered 1, 2, 3, ..., such that you can always get from one to the next.
2. Suppose also that, if the nth statement is true you can prove, by ordinary logical argument, that the $(n+1)$th statement is also true.
3. In that case, if the first statement is true the second must also be true,

and hence the third, and hence the fourth, and so on, for any value of n, however large.

This proves the statement to be true in the general case.

Example 3(i) Prove that the sum of the first n integers is $n(n+1)/2$.

Answer: Write the sum of the first n integers as $\sum_{i=1}^{i=n} i$. ($\sum_{i=a}^{i=b}$ means 'the sum of all terms like the one shown, with i taking all integer values from a to b'. Sometimes, when the limits are obvious, or have previously been specified, either or both of the indications $i=a$ and $i=b$ may be omitted; or they may be abbreviated, for example to meaning 'the summation of terms with all allowable, or all obvious, values of i'.) If the formula is true when $n=k$, then

$$\sum_{i=1}^{i=k} i = \frac{k(k+1)}{2}.$$

Hence,

$$\sum_{i=1}^{i=k+1} i = \left(\sum_{i=1}^{i=k} i\right) + (k+1)$$

$$= \frac{k(k+1)}{2} + (k+1)$$

$$= \frac{(k+1)(k+2)}{2}.$$

Thus if the formula is true for $n=k$ then it is also true for $n=k+1$. But it is true for $n=1$, since $1 \cdot 2/2 = 1$. Therefore it is true for $n=2$. (Check: $2 \cdot 3/2 = 3 = 1+2$.) Therefore it is true for $n=3$, and so on, for all values of n.

Example 3(j) Prove that R_n, where $R_n = 7^n - 1$ and n is an integer, is a multiple of 6.

Answer: Suppose the statement is true for $n=k$. Thus we can write $R_k = 6p$, where p is an integer. That is,

$$7^k - 1 = 6p.$$

$$\therefore \quad 7(7^k - 1) = 7(6p) = 42p.$$

$$\therefore \quad 7^{k+1} - 7 = 42p.$$

Hence,

$$R_{k+1} = 7^{k+1} - 1 = 42p + 6 = 6(7p+1).$$

So if R_k is a multiple of 6, R_{k+1} is also a multiple of 6. But the statement is true for $n=1$ ($R_1=6$), so it is also true for $n=2$ (check $R_2=48$), and hence for $n=2, 3, 4$, and all values of n.

EXERCISES 3C

1. Prove that $1^2+2^2+3^2+\cdots+n^2=n(n+1)(2n+1)/6$.
2. Prove that n^3-n is divisible by 3.

3.7 THE BINOMIAL THEOREM

The evaluation of an operation similar to $(a+bz)^p$, where a, b, z, and p are numbers, is one of the most frequent requirements in mathematics. To obtain either the exact result or an approximation to it, use is made of the binomial theorem which I shall discuss here, though not attempt to prove in detail. But first we can note that $(a+bz)=a(1+bz/a)=a(1+x)$ if we write x in place of bz/a. Since now $(a+bz)^p=(a(1+x))^p=a^p(1+x)^p$, and since a^p causes no trouble, we need worry only about $(1+x)^p$.

First consider the case where p is a positive integer. When $p=2$, there is no difficulty in working out the result:

$$\begin{aligned}(1+x)^2&=(1+x)(1+x)=1\cdot(1+x)+x\cdot(1+x)\\&=1+x+x+x^2=1+2x+x^2.\end{aligned}$$

Similarly,

$$\begin{aligned}(1+x)^3&=(1+x)(1+x)^2=(1+x)(1+2x+x^2)\\&=(1+2x+x^2)+(x+2x^2+x^3)\\&=1+3x+3x^2+x^3.\end{aligned}$$

And again,

$$(1+x)^4=1+4x+6x^2+4x^3+x^4,$$

and

$$(1+x)^5=1+5x+10x^2+10x^3+5x^4+x^5.$$

However, it would be too tedious to work through to the answer when $p=n$ and n has a large integer value. But to get quickly to that position we can *guess* at the answer by looking at the pattern of the expansions so far obtained, and then prove, by mathematical induction, that the guess gave the correct result.

Writing out just the coefficients of x^0, x^1, x^2, x^3, etc. in the expansion of $(1+x)^p$ we can set up the following scheme:

$$
\begin{array}{lccccccccccclcccccccccccc}
(1+x)^1, & & & & & 1 & & 1 & & & & & & & & & & {}^1C_0 & & {}^1C_1 & & & & & \\
(1+x)^2, & & & & 1 & & 2 & & 1 & & & & & & & & {}^2C_0 & & {}^2C_1 & & {}^2C_2 & & & & \\
(1+x)^3, & & & 1 & & 3 & & 3 & & 1 & & & & & & {}^3C_0 & & {}^3C_1 & & {}^3C_2 & & {}^3C_3 & & & \\
(1+x)^4, & & 1 & & 4 & & 6 & & 4 & & 1 & & & & {}^4C_0 & & {}^4C_1 & & {}^4C_2 & & {}^4C_3 & & {}^4C_4 & & \\
(1+x)^5, & 1 & & 5 & & 10 & & 10 & & 5 & & 1 & & {}^5C_0 & & {}^5C_1 & & {}^5C_2 & & {}^5C_3 & & {}^5C_4 & & {}^5C_5 \\
\vdots & & & & & & \vdots & & & & & & & & & & & & \vdots & & & & & &
\end{array}
$$

and so on.

In the triangular array of numbers in the middle, each number in any line is the coefficient of a power of x in the expansion of the term on the left. But also, each number is the sum of the two numbers immediately to the left and right of it in the line above. For example, the line with $(1+x)^4$ contains the coefficients $1=0+1$, $4=1+3$, $6=3+3$, $4=3+1$, and $1=1+0$. This triangular pattern of numbers is called Pascal's triangle. To the right of Pascal's triangle is another triangle in which numbers have been replaced by the symbols for appropriate combinations. Use of equation (3.4) shows that the value of the symbol in the 'combinations triangle' is the same as the number in the corresponding position in Pascal's triangle, e.g., ${}^3C_1=3$, ${}^4C_2=6$, etc.

In this situation we can guess that the coefficients in the expansion of $(1+x)^6$ will be 1, 6, 15, 20, 15, 6, and 1, to be rewritten as ${}^6C_0$, ${}^6C_1$, ${}^6C_2$, ${}^6C_3$, ${}^6C_4$, ${}^6C_5$, and ${}^6C_6$; and in the expansion of $(1+x)^7$ will be 1, 7, 21, 35, 35, 21, 7 and 1, to be written ${}^7C_0$, ${}^7C_1$, ${}^7C_2$, ${}^7C_3$, ${}^7C_4$, ${}^7C_5$, ${}^7C_6$, and ${}^7C_7$; and so on. So now at last there is an indication for a general formula for the expansion of $(1+x)^n$:

$$(1+x)^n={}^nC_0+{}^nC_1x+{}^nC_2x^2+{}^nC_3x^3+\cdots+{}^nC_nx^n, \tag{3.5}$$

that's a formula, clearly enough; but is it correct? We can prove the correctness by mathematical induction.

Suppose that equation (3.5) is true; then in the expansion of $(1+x)^n$ the coefficient of x^r is ${}^nC_r$. Now multiply both sides of equation (3.5) by $(1+x)$. It follows that in the expansion of $(1+x)^{n+1}$ the coefficient of x^r is ${}^nC_r+{}^nC_{r-1}$, whatever the value of r. But

$$
\begin{aligned}
{}^nC_r+{}^nC_{r-1} &= \frac{n!}{r!(n-r)!}+\frac{n!}{(r-1)!(n-r+1)!} \\
&= \frac{n!}{r!(n-r+1)!}[(n-r+1)+r] \\
&= \frac{(n+1)!}{r!(n+1-r)!} \\
&= {}^{n+1}C_r.
\end{aligned}
$$

Thus for any value of r, if the coefficient of x^r in the expansion of $(1+x)^n$ is ${}^nC_r$, then the coefficient of x^r in the expansion of $(1+x)^{n+1}$ is ${}^{n+1}C_r$. But the first part of the statement is certainly true for $n=1$ and also, indeed, for $n=2$, 3, 4, and 5 as demonstrated in the construction above of Pascal's triangle. Hence, by the argument of mathematical induction, it is true for all values of n; and so equation (3.5) is also true.

The formula of equation (3.5) applies to $(1+x)^n$ where n is a positive integer. Even if n is very large, the series on the r.h.s. of equation (3.5) eventually comes to an end because the coefficients of the successive powers of x, with the form $n(n-1)(n-2)(n-3)\ldots$ will, after the term in x^n, contain the factor $(n-n)$ and so will thereafter be zero.

Up to this point we have dealt with $(1+x)$ raised to positive integer powers. We can now consider the situation when the power is not a positive integer. The importance and strength of the binomial theorem lies in the fact, which can be proved but will not be proved here, that *whatever the value of p*, and provided x lies between -1 and $+1$,

$$(1+x)^p = 1 + px + \frac{p(p-1)}{1\cdot 2}x^2 + \frac{p(p-1)(p-2)}{1\cdot 2\cdot 3}x^3 + \cdots \tag{3.6}$$

The expansion of $(1+x)^p$ as given by the r.h.s. of equation (3.6) is a series in which the successive terms have exactly the same appearance or pattern as those when the power is an integer. However, the expansion is now an endless series if p is not a positive integer because in that case, however far one goes, the coefficients of the powers of x never contain a zero factor. Nevertheless, the successive terms in the series get smaller and smaller at a sufficient rate for the series to have a definite finite sum, equal to the left-hand side (l.h.s.) of equation (3.6). But, for unrestricted values of p, this is true only for values of x lying between -1 and $+1$, which can be written $-1<x<1$ (i.e. -1 is less than x which in turn is less than 1). This restriction on x must be adhered to if nonsense is to be avoided. (Equation (3.6) is also true for $x=1$ if and only if $p>-1$ and it is true for $x=-1$ if and only if $p \geqslant 0$.)

Example 3(k) Which is the greatest term numerically in $(1+x)^{20}$ if x has the value $\frac{4}{9}$? (Problems of this type arise in calculating the most probable outcome of events governed by a binomial probability distribution as discussed in Chapter 13.)

Answer: The $(r+1)$th term is $u_{r+1} = {}^{20}C_r x^r$. Where u_r is the rth term, $u_{r+1} > u_r$ provided that ${}^{20}C_r x^r > {}^{20}C_{r-1} x^{r-1}$. Since ${}^{20}C_r = (20-r+1)/r^{-1} \cdot {}^{20}C_{r-1}$ the condition for $u_{r+1} > u_r$ becomes $(21-r)/r^{-1} \cdot x > 1$, which in turn leads to $21x - rx > r$. Now substituting the value of $x = \frac{4}{9}$ gives $(84/9) - (4r/9) > r$ which in turn leads to the requirement that $r < 84/13$, i.e.

$r<6\frac{6}{13}$. Of course there is no term $u_{\frac{84}{13}+1}$, but the nearest to that figure when $r=6$ is u_7. In fact $u_7={}^{20}C_6(\frac{4}{9})^6=298.73$, while $u_6=268.86$ and $u_8=265.54$; and $(1\frac{4}{9})^{20}=1563.21$.

Example 3(l) Expand $(1-2x)^{\frac{3}{4}}$ and show that, if $x<0.3$, the error in terminating the expansion with the term in x^4 is less than 0.5%.

Answer: The wording of the question suggests that it is sufficient to expand as far as x^5 and then to show that the contribution of the last term is insignificant when x is not greater than 0.3.

$$(1-2x)^{\frac{3}{4}}=1+\tfrac{3}{4}(-2x)+\frac{\frac{3}{4}(\frac{3}{4}-1)}{1\cdot 2}(-2x)^2+\frac{\frac{3}{4}(\frac{3}{4}-1)(\frac{3}{4}-2)}{1\cdot 2\cdot 3}(-2x)^3$$

$$+\frac{\frac{3}{4}(\frac{3}{4}-1)(\frac{3}{4}-2)(\frac{3}{4}-3)}{1\cdot 2\cdot 3\cdot 4}(-2x)^4+\cdots$$

$$=1-\tfrac{3}{2}x-\tfrac{3}{8}x^2-\tfrac{5}{16}x^3-\tfrac{45}{128}x^4-\tfrac{117}{256}x^5-\cdots$$

When $x=0.3$, the r.h.s. on inclusion of successive terms has the values 1, 0.55000, 0.51625, 0.50781, 0.50496, 0.50385,... Evidently the term in x^5 makes a very small change to the sum, and the change would be even less for $x<0.3$. Calculation of $(1-2\times 0.3)^{\frac{3}{4}}$ gives 0.50297 and the sum up to x^4 (with $x=0.3$) is within 0.3% of this.

Example 3(m) Expand $(1+x-3x^2)^{-2}$ up to the term in x^3.

Answer: Write the expression as $y=(1+(x-3x^2))^{-2}$ and assume that $-1<(x-3x^2)<1$. Then

$$y=1+\frac{-2}{1}(x-3x^2)+\frac{-2\cdot -3}{1\cdot 2}(x-3x^2)^2+\frac{-2\cdot -3\cdot -4}{1\cdot 2\cdot 3}(x-3x^2)^3+\cdots$$

$$=1-2(x-3x^2)+3(x^2-6x^3+9x^4)-4(x^3-9x^4+27x^5-27x^6)+\cdots$$

$$=1-2x+9x^2-22x^3+\cdots$$

where the last line is obtained by picking out the appropriate part of each bracket.

EXERCISES 3D

1. Find the largest term of $(p+q)^{16}$ when $p=0.6$ and $q=0.4$. (Hint: $(p+q)^{16}=P(1+s)^{16}$ where $P=p^{16}$ is a constant quantity multiplying every term of the expansion and $s=q/p$.) NB: in a real situation, p and q

might be the probabilities of success and failure in a test repeated 16 times. If so, what is the interpretation of the answer?

2. Provide the first five terms of the expansions of
   (a) $(1+x)^{\frac{1}{2}}$;
   (b) $(1+3y^2)^{-2}$;
   (c) $(1-z)^{-1}$.
3. Show that the first four terms of the expansion of $\sqrt{(1-x)/(1+x)}$ are given by $1-x+x^2/2-x^3/2$ and hence that this can be used to evaluate the expression to better than 2 parts in 5000 for $0<x<0.15$.

3.8 THE EXPONENTIAL FUNCTION

If there is a number E defined by

$$E=\left(1+\frac{1}{n}\right)^{nx}$$

for some values of n and x, its expansion is given by

$$E=1+nx\cdot\frac{1}{n}+\frac{nx(nx-1)}{1\cdot 2}\left(\frac{1}{n}\right)^2+\frac{nx(nx-1)(nx-2)}{1\cdot 2\cdot 3}\left(\frac{1}{n}\right)^3+\cdots$$

$$=1+x+\frac{x\left(x-\frac{1}{n}\right)}{1\cdot 2}+\frac{x\left(x-\frac{1}{n}\right)\left(x-\frac{2}{n}\right)}{1\cdot 2\cdot 3}+\cdots$$

Whatever the value of x, other than zero, as n becomes very very large, a term of the kind $(x-k/n)$ becomes only infinitesimally different from x. In fact, for n large enough, the difference is less than ε, however small we choose to make ε. The process of letting n become indefinitely large is called 'letting n tend to infinity'. This is a deceptive form of words, because there is no real number 'infinity' towards which the number n tends. The situation is very different from saying that, as n becomes indefinitely large, $1/n$ tends to zero; zero is a number we deal with regularly. Although it would be better (in the sense of providing fewer philosophical problems) to let $1/n$ tend to zero, the usage of letting n tend to infinity, written as $n\to\infty$, is firmly established and will appear several times in later chapters. In the present case, it can be shown that, as n 'tends to infinity', E tends to a definite quantity, called its limit, which is written as e^x. The notation is $\lim_{n\to\infty} E$, which is read as 'the limit of E as n tends to infinity'.

So,

$$\lim_{n\to\infty} E=1+x+\frac{x^2}{2!}+\frac{x^3}{3!}+\frac{x^4}{4!}+\cdots=e^x. \tag{3.7}$$

Since

$$\left(1+\frac{1}{n}\right)^{nx}=\left(\left(1+\frac{1}{n}\right)^{n}\right)^{x}$$

(by Section 3.2) it can be readily understood that

$$\begin{aligned} e &= \lim_{n\to\infty} E \qquad \text{when } x=1 \\ &= \lim_{n\to\infty}\left(1+\frac{1}{n}\right)^{n} \\ &= 1+1+1/2!+1/3!+1/4!+\cdots \\ &= 2.71828\ldots \end{aligned}$$

The quantity e, the exponential number, is a number of fundamental importance in mathematics, similar in its ubiquitous appearance to π. It is the base of natural logarithms because of the behaviour of e^x in the differential and integral calculus (cf. Section 3.3 and Chapter 10). The expansion of e^x given by equation (3.7) is also a most important relation which you will frequently meet later in this book.

4

Trigonometry

Trigonometry is defined in the Oxford English Dictionary as the 'branch of mathematics which deals with the measurement of the sides and angles of triangles and with certain functions of their angles or of angles in general ...and ... including the theory of triangles, of angles, and of (elementary) singly periodic functions'. I can't say that triangles never appear in the biological sciences, but if they do, it's not very often. However, periodicity, either in time or in space, is ubiquitous in the biological as in the other sciences, and the simplest way of studying periodicity is by means of the (elementary) singly periodic functions, namely the functions which you have surely met before – sine, cosine, tangent, etc. It is thus just the last part of dictionary-defined trigonometry that we're after, although we must pass through some corners of the other parts to get there.

4.1 ANGLES

Degrees are like metres or seconds or amps: they have dimensions in the physics sense of being measures in calibrated and empirically chosen units (this being the case both for degrees of angular measure and degrees of temperature measure). A problem with measures, sometimes called dimensioned numbers, is that there may be restrictions on the functions of which they might be the argument. One could deal, for example, with square metres or cubic metres, but the ideas of square root metres or logarithmic volts are without meaning. In the case of angles, it turns out to be particularly useful to have a measure which is independent of calibrated units. This is done by making the measure, called a radian, a *ratio* of two lengths. Being a ratio, it is dimensionless. The measure of an angle subtended at the centre of a circle by an arc is defined as the ratio of the length of the arc to the radius of the circle. So in Fig. 4.1 the angle θ (NB: it is common in mathematical texts to use Greek letters for angles) is

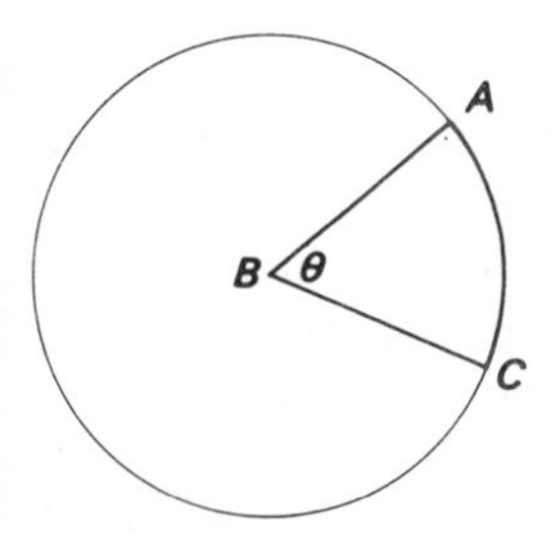

Figure 4.1

measured in radians by the ratio of the length of arc CA to radius AB.

By convention, the standard direction of rotation in mathematical work is anticlockwise. So in Fig. 4.1 the angle θ is swept out in proceeding round the circle from C to A. If the direction of travel were from A to C the angle swept out would be $-\theta$. A reference to direction in this way can be used to give a meaning to negative angles, although usually the matter is of no importance and an angle is assumed to be positive.

The circumference of the whole circle is given by the well-known formula $2\pi r$, so the complete angle, 'right round', at the centre has the value $2\pi r/r$, i.e. 2π radians. Thus 2π radians corresponds to 360°. A right angle is subtended at the centre by a quarter of the circumference, so $\pi/2$ radians corresponds to 90°. Table 4.1 lists some correspondences for your convenience. In the remainder of this book, angles are always expressed in radians unless it is explicitly stated otherwise.

Apropos of angles, you no doubt remember that in any triangle the sum of all three angles is π (=180°). In the case of a triangle containing a right angle (i.e. an angle of size $\pi/2$) the sum of the two acute angles is $\pi/2$. And you will also remember the theorem of Pythagoras, that in such a right-angled triangle the square on the hypotenuse equals the sum of the squares on the other two sides; e.g. with reference to the triangle of Fig. 4.2, $AB^2 = BC^2 + CA^2$. (Strictly, these statements refer only to planar triangles, i.e. those drawn on a flat surface; they are not true for triangles drawn on the surface of a sphere, as you can see by examining the triangles formed by the equator and lines of longitude separated by 90° on a terrestrial globe.)

Table 4.1 Correspondence between radians and degrees as measures of angles

radians	2π	π	$\pi/2$	$\pi/3$	$\pi/4$	$\pi/6$	1	0.01745
degrees	360	180	90	60	45	30	57.296	1

4.2 THE CIRCULAR FUNCTIONS

The simple circular functions are sine (sin), cosine (cos) and tangent (tan). They are the '(elementary) singly periodic functions' mentioned in the dictionary and their definitions seem to be one of the few bits of post-sixteenth century mathematics commonly remembered. The definitions are nevertheless repeated here for completeness. Referring to triangle ABC of Fig. 4.2, where angle $\angle ABC$ is θ and angle $\angle ACB$ is a right angle (and so has the value $\pi/2$ or $90°$), the side AB is the hypotenuse. (NB: this word occurs only with respect to triangles containing a right angle.) Call AC the opposite side and BC the adjacent side, i.e. opposite and adjacent to the angle θ. Note also that the other acute angle $\angle BAC = \pi/2 - \theta$. Then the sine of an angle is defined as the ratio of the lengths of the opposite side to the hypotenuse, the cosine as the adjacent side to the hypotenuse, and the tangent as the opposite side to the adjacent side.
Thus

$$\sin\theta = \frac{AC}{AB} = \frac{\text{opposite}}{\text{hypotenuse}} \quad \text{and} \quad \sin\theta = \cos\left(\frac{\pi}{2} - \theta\right)$$

$$\cos\theta = \frac{BC}{AB} = \frac{\text{adjacent}}{\text{hypotenuse}} \quad \text{and} \quad \cos\theta = \sin\left(\frac{\pi}{2} - \theta\right)$$

$$\tan\theta = \frac{AC}{BC} = \frac{\text{opposite}}{\text{adjacent}} \quad \text{and} \quad \tan\theta = \cot\left(\frac{\pi}{2} - \theta\right).$$

From these definitions it follows that $\tan\theta = (\sin\theta/\cos\theta)$ since

$$\frac{AC}{AB}\bigg/\frac{BC}{AB} = \frac{AC}{BC}.$$

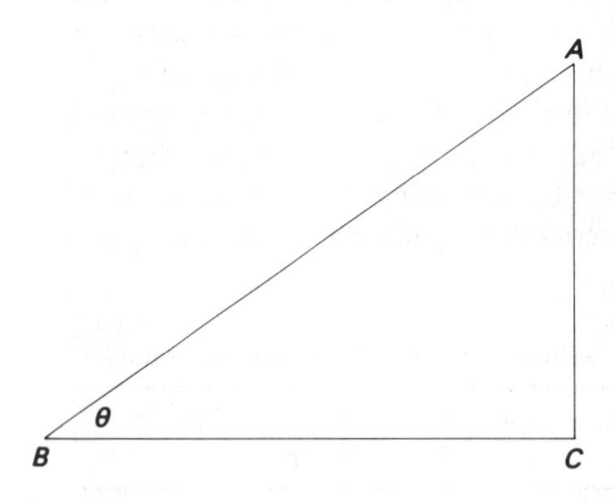

Figure 4.2

The reciprocals of these functions are often encountered. They are

secant (sec), the reciprocal of cosine, so that $\sec\theta=\dfrac{AB}{BC}$,

cosecant (cosec), the reciprocal of sine, so that $\operatorname{cosec}\theta=\dfrac{AB}{AC}$, and

cotangent (cot), the reciprocal of tangent, so that $\cot\theta=\dfrac{BC}{AC}$.

These definitions give the circular functions meaning only for angles less than $\pi/2$. We certainly need to deal with the functions of larger angles so that wider definitions are needed and we shall meet these shortly.

4.3 SOME RELATIONS BETWEEN THE CIRCULAR FUNCTIONS

For the triangle in Fig. 4.2 we have by Pythagoras that $BC^2+CA^2=AB^2$. So dividing through by AB^2 we have

$$\frac{BC^2}{AB^2}+\frac{CA^2}{AB^2}=1, \qquad \text{i.e. } (\cos\theta)^2+(\sin\theta)^2=1.$$

This is usually written

$$\cos^2\theta+\sin^2\theta=1. \tag{4.1}$$

Since $\tan\theta=(\sin\theta)/(\cos\theta)$, therefore $\tan^2\theta=(\sin^2\theta)/(\cos^2\theta)$ and from equation (4.1) $\tan^2\theta=(1-\cos^2\theta)/(\cos^2\theta)$. Hence,

$$\tan^2\theta=\sec^2\theta-1. \tag{4.2}$$

Similarly,

$$\cot^2\theta=\operatorname{cosec}^2\theta-1. \tag{4.3}$$

Statements like those of equations (4.1), (4.2) and (4.3) are true for any value of the angle θ. They should properly be called identities rather than equations, implying that they remain identically true for all values of the angle (cf. Chapter 7).

Example 4(a) Show that

$$\frac{(\sec\theta-1)(\cos\theta+1)}{\cos\theta}=\tan^2\theta.$$

Answer:

$$\frac{(\sec\theta-1)(\cos\theta+1)}{\cos\theta}=\frac{\sec\theta\cos\theta+\sec\theta-\cos\theta-1}{\cos\theta}.$$

But $\sec\theta=1/(\cos\theta)$, so the r.h.s. is

$$\frac{1+\sec\theta-\cos\theta-1}{\cos\theta}=\sec^2\theta-1$$

$$=\tan^2\theta \qquad \text{from equation (4.2).}$$

Example 4(b) Show that

$$\frac{1+\cos\theta}{\sin\theta}=\frac{\sin\theta}{1-\cos\theta}.$$

Answer:

$$\frac{1+\cos\theta}{\sin\theta}-\frac{\sin\theta}{1-\cos\theta}=\frac{(1+\cos\theta)(1-\cos\theta)-\sin^2\theta}{\sin\theta(1-\cos\theta)}$$

$$=\frac{1-\cos^2\theta-\sin^2\theta}{\sin\theta(1-\cos\theta)}$$

$$=0.$$

$$\therefore \qquad \frac{1+\cos\theta}{\sin\theta}=\frac{\sin\theta}{1-\cos\theta}.$$

EXERCISES 4A

1. Show that

   (a) $\tan\theta+\cot\theta=\sec\theta\cdot\operatorname{cosec}\theta$;

   (b) $\dfrac{\sin\theta}{1+\cos\theta}=\operatorname{cosec}\theta-\cot\theta.$

2. Show that

$$\frac{\sin^4\theta+\cos^4\theta}{\sin\theta\cdot\cos\theta}=\sec\theta\operatorname{cosec}\theta-2\sin\theta\cos\theta.$$

4.4 VALUES OF THE CIRCULAR FUNCTIONS FOR ANGLES GREATER THAN $\pi/2$

Imagine, as in Fig. 4.3, that in a circle with centre B the radius BA sweeps out a steadily increasing angle θ, that the radius is always counted as being

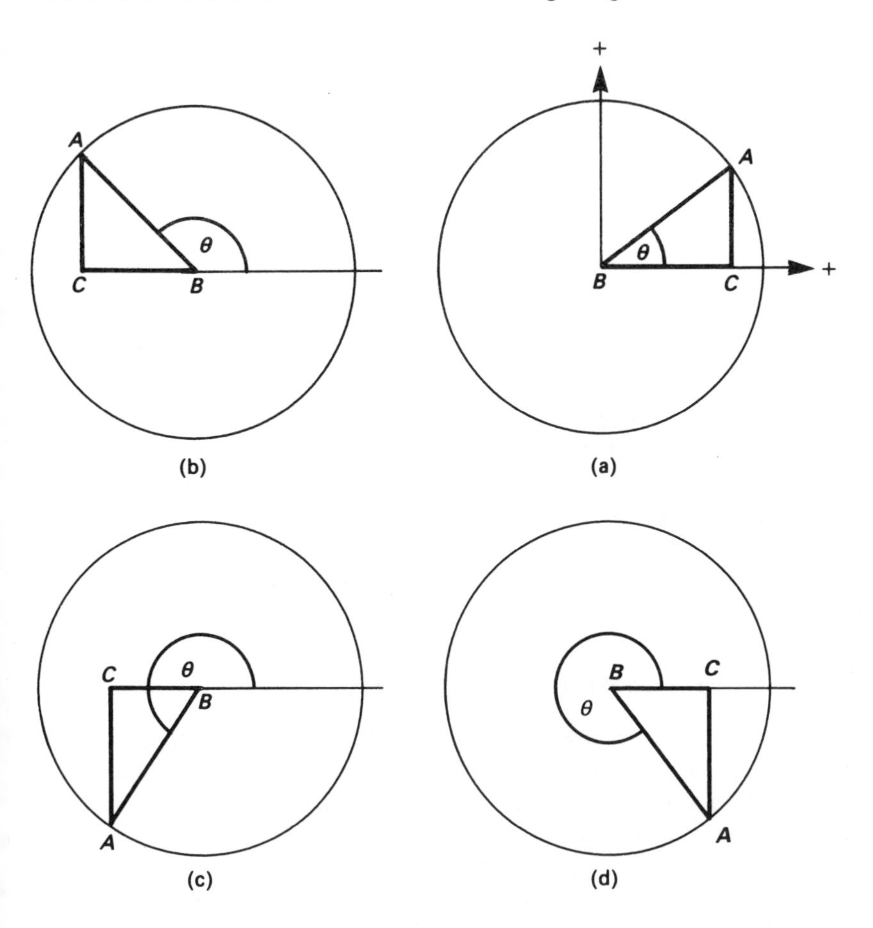

Figure 4.3

positive and that directions up from B and to the right of B are regarded as positive and those to the left of and down from B are regarded as negative. In all the cases shown in Fig. 4.3 we have $\sin\theta = AC/AB$ and AB is positive. In the first quadrant (Fig. 4.3 (a)), with θ taking any value between 0 and $\pi/2$, written as $0 \leqslant \theta \leqslant \pi/2$, AC is positive (+ve) and so $\sin\theta$ is +ve. In the second quadrant (Fig. 4.3 (b)), $\pi/2 \leqslant \theta \leqslant \pi$, AC is still +ve, so $\sin\theta$ remains +ve. In the third quadrant (Fig. 4.3 (c)), $\pi \leqslant \theta \leqslant 3\pi/2$, AC is negative (−ve) and so $\sin\theta$ is −ve. Finally, in the fourth quadrant (Fig. 4.3 (d)), $3\pi/2 \leqslant \theta \leqslant 2\pi$, AC remains −ve and so does $\sin\theta$.

The other functions for angles greater than $\pi/2$ can be teased out in the same way. For $\cos\theta$ given by BC/AB, it follows that $\cos\theta$ is +ve in the first quadrant, −ve in the second and third, and +ve again in the fourth. And $\tan\theta$ is +ve in the first quadrant, −ve in the second, +ve again in the third,

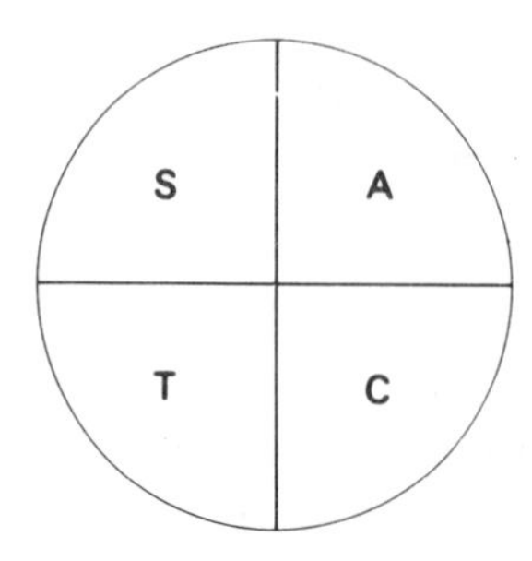

Figure 4.4

and −ve again in the fourth. The signs of the sec function are the same as those of cos, of the cosec function as those of sin, and of the cot function as those of tan. In Fig. 4.4 the letters indicate which functions are +ve in the various quadrants: starting at the lower right (fourth) quadrant and going round anticlockwise, the +ve functions are C (=cos), A (=all), S (=sin) and T(=tan). The mnemonic for remembering this is the word CAST, provided that you can remember at which quadrant to begin and in which direction to go.

4.5 INVERSE CIRCULAR FUNCTIONS

Note that if $x=\sin\theta$, then sin is a function which maps from the domain of θ to the range of x and θ is the argument of the function. The superscript $^{-1}$ following a variable which stands for a number, such as x or θ in the previous sentence, provides the reciprocal of the variable, i.e. $x^{-1}=1/x$. But the same superscript applied to a function provides the *inverse function.* If $x=\sin\theta$, then $\theta=\sin^{-1}x$, which is read as 'θ equals sine to the minus one of x' and which means 'θ is the angle of which the sine is x'. Thus the $\sin^{-1}$ function maps from the domain of x to the range of θ. The inverse functions $\cos^{-1}$, $\tan^{-1}$, etc. can be defined in ways corresponding to the definition of $\sin^{-1}$. For the functions $\sin^{-1}$ and $\tan^{-1}$ it is to be understood that, to avoid ambiguity, the resulting angle lies in the range $-\pi/2$ to $\pi/2$, whereas for $\cos^{-1}$ the range is 0 to π. The function $\sin^{-1}x$ is sometimes symbolized by arcsin x and similarly for $\cos^{-1}x$, $\tan^{-1}x$, etc.

But watch out for a trap. If, once again, $x=\sin\theta$, then

$$(\sin\theta)^{-1}=x^{-1}=\frac{1}{x}=\operatorname{cosec}\theta.$$

4.6 VALUES OF THE CIRCULAR FUNCTIONS FOR KEY ANGLES

In Fig. 4.3 (a), when A is very close to C, AC is effectively zero and AB is effectively equal to BC. So, from the first definitions of the functions,

$\sin 0=0$, $\cos 0=1$ and $\tan 0=0$. Note also that, when θ is very small, but not quite zero, AC/AB in Fig. 4.3(a) is very nearly the same as the definition of the measure of θ in radians as illustrated in Fig. 4.1 where $\theta=AC/AB$. So for very small angles, $\cos\theta\approx 1$ and $\sin\theta\approx\theta$, but only if the angle is measured in radians.

As θ approaches $\pi/2$, AC becomes effectively equal to AB while BC falls effectively to zero. Hence $\sin(\pi/2)=1$, $\cos(\pi/2)=0$ and $\tan(\pi/2)$, the calculation of which now involves division by zero, is undefined, although it is sometimes said that $\tan\theta$ tends to infinity as θ tends to $\pi/2$. Note that if θ tends towards $\pi/2$ from below (i.e. if θ is in the first quadrant) $\tan\theta$ is positive and so would be tending to $+\infty$; whereas if θ tends towards $\pi/2$ from above (i.e. if θ is in the second quadrant) $\tan\theta$ is negative and so would be tending to $-\infty$. This illustrates why $\tan(\pi/2)$ is undefined. But while $\tan\theta$ can take very large positive or negative values, the magnitudes of $\cos\theta$ and $\sin\theta$ (i.e. their absolute values, ignoring the + or − signs) are never greater than 1. That is, if $y=\sin x$, the function sin maps from the domain of x which extends over all real numbers, however large or small, to a range of y which extends only over real numbers between $+1$ and -1.

In the right-angled triangle of Fig. 4.5, suppose that AC and BC are both of length 1. Then by Pythagoras, $AB^2=BC^2+AC^2=2$ so that AB is of length $\sqrt{2}$. But also, ABC is a isosceles triangle and since the sum of the angles in a triangle is π, it follows that the angles $\angle ABC=\angle BAC=\pi/4$ $(=45°)$. Hence

$$\sin\frac{\pi}{4}=\cos\frac{\pi}{4}=\frac{1}{\sqrt{2}}=0.707 \qquad \text{and} \qquad \tan\frac{\pi}{4}=1.$$

In the triangle PQR of Fig. 4.6, suppose the three sides are equal, each of length 2, in which case the three angles will also be equal, each of value

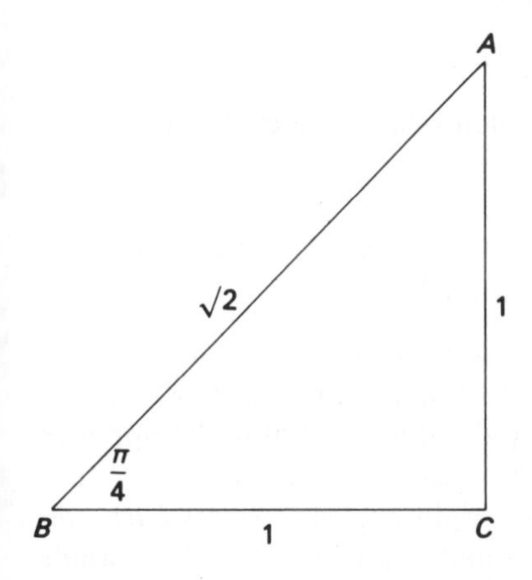

Figure 4.5

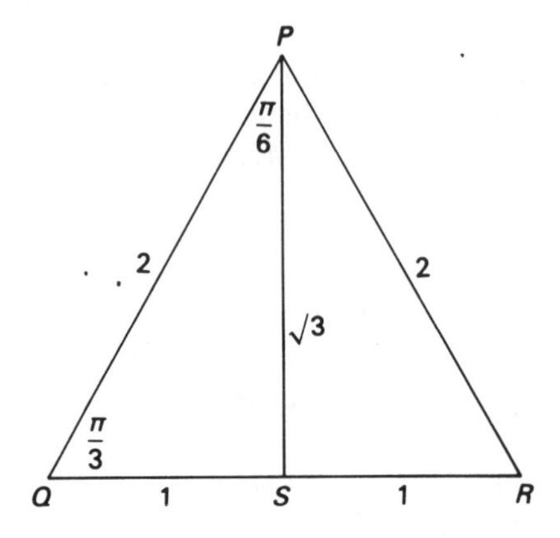

Figure 4.6

$\pi/3$ ($=60°$). Let PS be the perpendicular bisector of the side QR; then in the right-angled triangle PQS, PQ is of length 2, QS is of length 1 and, by Pythagoras, PS is of length $\sqrt{3}$. But also, PS bisects the angle $\angle QPR$ so that in the triangle PQS the angle $\angle PQS = \pi/3$ ($=60°$) and the angle $\angle QPS = \pi/6$ ($=30°$). Hence it follows that

$$\sin \pi/6 = \cos \pi/3 = \frac{1}{2} = 0.500$$

$$\sin \pi/3 = \cos \pi/6 = \frac{\sqrt{3}}{2} = 0.866$$

$$\tan \pi/6 = \frac{1}{\sqrt{3}} = 0.577$$

$$\tan \pi/3 = \sqrt{3} = 1.732.$$

The various values established here are collected together in Table 4.2.

4.7 THE TRIGONOMETRIC FUNCTIONS OF SUMS AND DIFFERENCES OF ANGLES

The construction in Fig. 4.7 is drawn to allow calculation of the sine and cosine of the sum of the two angles $\angle DCG$ and $\angle ACD$, marked θ and ϕ, respectively. All the angles marked ⌈ are right angles. Because the opposite angles in a pair of intersecting lines are equal (specifically $\angle CFB$ and $\angle DFA$, marked •), it follows that in the two right-angled triangles CBF and ADF, the angles $\angle BCF$ and $\angle DAF$ are equal and of value θ. Also, angle

Table 4.2 Values of trigonometric functions of some angles

angle	sin	cos	tan
0	0	1	0
$\pi/6=30°$	$\frac{1}{2}$	$\frac{\sqrt{3}}{2}$	$\frac{1}{\sqrt{3}}$
$\pi/4=45°$	$\frac{1}{\sqrt{2}}$	$\frac{1}{\sqrt{2}}$	1
$\pi/3=60°$	$\frac{\sqrt{3}}{2}$	$\frac{1}{2}$	$\sqrt{3}$
$\pi/2=90°$	1	0	(not defined)

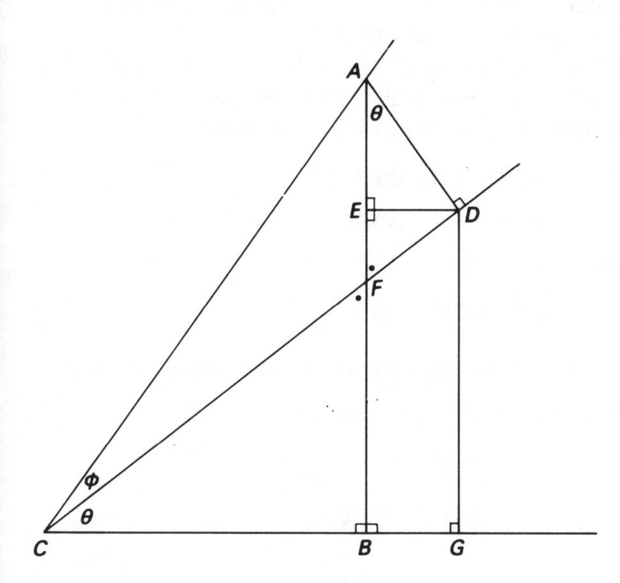

Figure 4.7

$\angle ACB=\theta+\phi$. Now

$$\sin(\theta+\phi)=\frac{BA}{AC}$$

$$=\frac{BE+EA}{AC} \qquad \text{since } BA=BE+EA$$

$$=\frac{GD}{AC}+\frac{EA}{AC} \qquad \text{since } BE=GD$$

$$=\frac{GD}{DC}\cdot\frac{DC}{AC}+\frac{EA}{DA}\cdot\frac{DA}{AC}$$

$$=\sin\theta\cos\phi+\cos\theta\sin\phi. \tag{4.4}$$

A similar argument leads to the result

$$\cos(\theta+\phi)=\cos\theta\cos\phi-\sin\theta\sin\phi. \tag{4.5}$$

There are corresponding formulae for the sine and cosine of $(\theta-\phi)$ viz.

$$\sin(\theta-\phi)=\sin\theta\cos\phi-\cos\theta\sin\phi \tag{4.6}$$

$$\cos(\theta-\phi)=\cos\theta\cos\phi+\sin\theta\sin\phi \tag{4.7}$$

[Note the difference of sign between the l.h.s. and r.h.s. in the formulae for the cosine of the compound angles, equations (4.5) and (4.7).]

It can be seen from Fig. 4.3 that $\sin(\theta+2\pi)=\sin\theta$. Alternatively, since $\sin 2\pi=\sin 0=0$ and $\cos 2\pi=\cos 0=1$, which may also be seen from Fig. 4.3, these results can be used in equations (4.4) and (4.5) to give

$$\sin(\theta+2\pi)=\sin\theta$$
$$\cos(\theta+2\pi)=\cos\theta,$$

from which it follows that

$$\tan(\theta+2\pi)=\tan\theta.$$

If the angles θ and ϕ are equal, we can write 2θ in place of $(\theta+\phi)$ and then

$$\sin 2\theta=2\sin\theta\cos\theta$$

and

$$\begin{aligned}\cos 2\theta&=\cos^2\theta-\sin^2\theta\\&=2\cos^2\theta-1\\&=1-2\sin^2\theta.\end{aligned}$$

If $\theta=\pi/2$ $(=90°)$ so that $2\theta=\pi$, it follows that $\sin\pi=0$ and $\cos\pi=-1$.

If $\phi=\pi/2$ when θ and ϕ are unequal, it follows that

$$\sin\left(\frac{\pi}{2}-\theta\right)=\cos\theta \qquad \text{and} \qquad \sin\left(\frac{\pi}{2}+\theta\right)=\cos\theta$$

$$\cos\left(\frac{\pi}{2}-\theta\right)=\sin\theta \qquad \text{and} \qquad \cos\left(\frac{\pi}{2}+\theta\right)=-\sin\theta.$$

Dividing the sine by the cosine of the respective angles,

$$\tan\left(\frac{\pi}{2}-\theta\right)=\cot\theta \qquad \text{and} \qquad \tan\left(\frac{\pi}{2}+\theta\right)=-\cot\theta.$$

Example 4(c) Show that $\sin 3A=3\sin A-4\sin^3 A$.

Answer:

$$\begin{aligned}\sin 3A&=\sin(2A+A)\\&=\sin 2A\cos A+\cos 2A\sin A\\&=2\sin A\cos^2 A+\sin A(\cos^2 A-\sin^2 A)\\&=3\sin A\cos^2 A-\sin^3 A\\&=3\sin A(1-\sin^2 A)-\sin^3 A\\&=3\sin A-4\sin^3 A.\end{aligned}$$

Example 4(d) Show that

$$\frac{\sin X}{\sin Y}+\frac{\cos X}{\cos Y}=\frac{2\sin(X+Y)}{\sin 2Y}.$$

Answer:

$$\begin{aligned}\frac{\sin X}{\sin Y}+\frac{\cos X}{\cos Y}&=\frac{\sin X\cos Y+\cos X\sin Y}{\sin Y\cos Y}\\&=\frac{2\sin(X+Y)}{\sin 2Y}.\end{aligned}$$

Example 4(e) Show that

$$\frac{\cot P-\tan Q}{1+\cot P\tan Q}=\cot(P+Q).$$

Answer: Multiply the l.h.s. by

$$\frac{\sin P\cos Q}{\sin P\cos Q}$$

giving

$$\begin{aligned}\frac{\cos P\cos Q-\sin P\sin Q}{\sin P\cos Q+\cos P\sin Q}&=\frac{\cos(P+Q)}{\sin(P+Q)}\\&=\cot(P+Q).\end{aligned}$$

EXERCISES 4B

1. Show that $\cos 3X = 4\cos^3 X - 3\cos X$ and check the statement for the case where $3X = \pi/2$.
2. Show that

$$\frac{1}{\cot\theta - \tan\theta} - \frac{1}{\cot\theta + \tan\theta} = \frac{\tan 2\theta}{\operatorname{cosec}^2\theta}.$$

4.8 SUMS AND DIFFERENCES OF SINES AND COSINES

Adding together equations (4.4) and (4.6) gives

$$\sin(\theta + \phi) + \sin(\theta - \phi) = 2\sin\theta\cos\phi.$$

Now put $\theta + \phi = A$ and $\theta - \phi = B$ so that $A + B = 2\theta$ (or $\theta = (A+B)/2$) and $A - B = 2\phi$ (or $\phi = (A-B)/2$). This provides the formula

$$\sin A + \sin B = 2\sin\frac{A+B}{2}\cos\frac{A-B}{2}. \tag{4.8}$$

There are corresponding formulae for the difference of two sines and for the sum and difference of two cosines and of two tangents. All of these, together with a few others which have not been derived or mentioned here, are collected in Table 4.3.

Example 4(f) Show that

$$\frac{\sin 7R + \sin 5R}{\cos 4R + \cos 2R} = 2\sin 3R.$$

Answer:

$$\frac{\sin 7R + \sin 5R}{\cos 4R + \cos 2R} = \frac{2\sin 6R\cos R}{2\cos 3R\cos R} = \frac{\sin 6R}{\cos 3R}$$

$$= \frac{2\sin 3R\cos 3R}{\cos 3R}$$

$$= 2\sin 3R.$$

Example 4(g) Calculate the value of $\tan \pi/8$.

Answer: From Table 4.3,

$$\tan 2A = \frac{2\tan A}{1 - \tan^2 A}.$$

Table 4.3 Useful trigonometric formulae

$$\sin\left(\frac{\pi}{2}-\theta\right)=\cos\theta \qquad \sin\left(\frac{\pi}{2}+\theta\right)=\cos\theta$$

$$\cos\left(\frac{\pi}{2}-\theta\right)=\sin\theta \qquad \cos\left(\frac{\pi}{2}+\theta\right)=-\sin\theta$$

$$\tan\left(\frac{\pi}{2}-\theta\right)=\cot\theta \qquad \tan\left(\frac{\pi}{2}+\theta\right)=-\cot\theta$$

$$\cot\left(\frac{\pi}{2}-\theta\right)=\tan\theta \qquad \cot\left(\frac{\pi}{2}+\theta\right)=-\tan\theta$$

$$\cos^2\theta+\sin^2\theta=1$$

$$\sec^2\theta-\tan^2\theta=1 \qquad \mathrm{cosec}^2\theta-\cot^2\theta=1$$

$$\sin(\theta+\phi)=\sin\theta\cos\phi+\cos\theta\sin\phi$$

$$\sin(\theta-\phi)=\sin\theta\cos\phi-\cos\theta\sin\phi$$

$$\cos(\theta+\phi)=\cos\theta\cos\phi-\sin\theta\sin\phi$$

$$\cos(\theta-\phi)=\cos\theta\cos\phi+\sin\theta\sin\phi$$

$$\tan(\theta+\phi)=(\tan\theta+\tan\phi)/(1-\tan\theta\tan\phi)$$

$$\tan(\theta-\phi)=(\tan\theta-\tan\phi)/(1+\tan\theta\tan\phi)$$

$$\sin 2\theta=2\sin\theta\cos\theta$$

$$\cos 2\theta=\cos^2\theta-\sin^2\theta=2\cos^2\theta-1=1-2\sin^2\theta$$

$$\tan 2\theta=2\tan\theta/(1-\tan^2\theta)$$

$$\sin\theta+\sin\phi=2\sin\frac{\theta+\phi}{2}\cos\frac{\theta-\phi}{2}$$

$$\sin\theta-\sin\phi=2\cos\frac{\theta+\phi}{2}\sin\frac{\theta-\phi}{2}$$

$$\cos\theta+\cos\phi=2\cos\frac{\theta+\phi}{2}\cos\frac{\theta-\phi}{2}$$

$$\cos\theta-\cos\phi=2\sin\frac{\theta+\phi}{2}\cos\frac{\phi-\theta}{2}$$

Now, $\pi/4=2\pi/8$ and from Table 4.2, $\tan \pi/4=1$. Put $t=\tan \pi/8$. Then

$$1=\frac{2t}{1-t^2}.$$

Hence $$t^2+2t-1=0.$$

$$\therefore \qquad t=\frac{-2\pm\sqrt{4+4}}{2}.$$

Therefore since t must be +ve, $t=\sqrt{2}-1\approx 0.414$.

EXERCISES 4C

1. Show that $\dfrac{\sin 6X+\sin 2X}{\cos 6X+\cos 2X}=\tan 4X$.

2. Calculate the values of
   (a) $\cos \pi/12$;
   (b) $\sin \pi/12$;
   (c) $\tan \pi/12$.

4.9 THE AUXILIARY ANGLE

When an expression of the form $y=a\sin\theta+b\cos\theta$ appears, it is sometimes convenient to manipulate it so that y becomes a function of θ containing just one periodic function instead of two. This can be accomplished by comparing the equation for y with the expression $\sin(\theta+\phi)=\sin\theta\cos\phi+\cos\theta\sin\phi$. If a were equal to $\cos\phi$ and b were to equal $\sin\phi$, there would be no problem, for y could then be written directly as the sine of the sum of two angles. However, there is no reason to expect $\phi=\cos^{-1}a=\sin^{-1}b$, and indeed a and b are quite likely to have values greater than 1 and so could not possibly be a sine and cosine, respectively. Nevertheless, the matter can be forced by the simple trick illustrated in Fig. 4.8. In a right-angled triangle with sides a and b (this can always be constructed) designate as ϕ the acute angle adjacent to the side of length a. Then by Pythagoras' theorem the length of the hypotenuse is $\sqrt{a^2+b^2}$. Now write

$$y=\sqrt{a^2+b^2}\left(\sin\theta\frac{a}{\sqrt{a^2+b^2}}+\cos\theta\frac{b}{\sqrt{a^2+b^2}}\right)$$

$$=\sqrt{a^2+b^2}\,(\sin\theta\cos\phi+\cos\theta\sin\phi)$$

$$=\sqrt{a^2+b^2}\,(\sin(\theta+\phi))$$

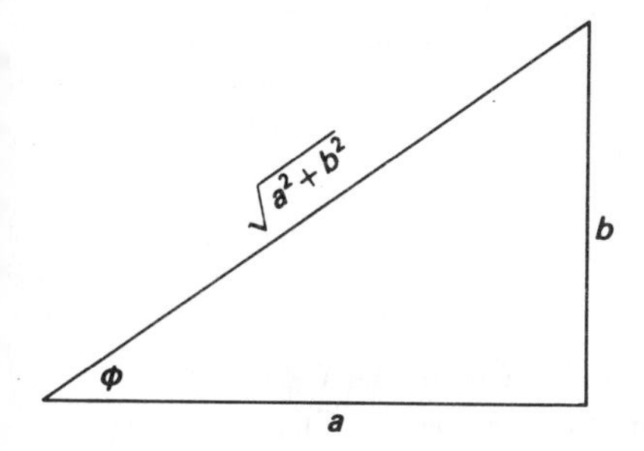

Figure 4.8

where $\cos\theta = a/\sqrt{a^2+b^2}$ and $\sin\phi = b/\sqrt{a^2+b^2}$ so that $\tan\phi = b/a$ or $\phi = \tan^{-1}b/a$. In these circumstances, ϕ is called the auxiliary angle in the function which yields y.

The specification of ϕ as $\tan^{-1}b/a$ is, however, still incomplete. This is because ϕ and $2n\pi + \phi$, for any integer value of n, have the same value of tangent (and also of sine and cosine). Hence to avoid $\tan^{-1}b/a$ having an unlimited number of possibilities, ϕ must be restricted and the restriction is $-\pi < \phi < \pi$.

In electronic and biological signal analysis and signal processing, the signal is often in the form of a wave or other disturbance of the form $A\sin(2\pi ft)$. The quantity A is the amplitude of the signal; the quantity in brackets is called its phase and it varies with time, t, at a frequency f. When two signals of the same frequency are mixed, the resultant is a signal $A'\sin(2\pi ft + \phi)$ and ϕ is sometimes called the phase shift of the new signal with respect to the original. Notice that when $\phi = \pi/2$ the new signal becomes $A'\cos(2\pi ft)$ since $\sin(\theta + \pi/2) = \cos\theta$; the new signal is the said to be '$\pi/2$ out of phase' with the original.

Example 4(h) Express $3\sin\theta + 4\cos\theta$ in the form $R\sin(\theta + \phi)$.

Answer: Since $\sqrt{3^2+4^2} = 5$, put $\cos\phi = \frac{3}{5}$ and $\sin\phi = \frac{4}{5}$.
Then $3\sin\theta + 4\cos\theta = 5\sin(\theta + \phi)$ where

$$\begin{aligned}\phi &= \tan^{-1}(\tfrac{4}{3})\\ &= 0.9273 \text{ radians}\\ &= 53.13°.\end{aligned}$$

Example 4(i) A radio wave $A\sin(2\pi ft)$ is mixed with 50% of the out-of-phase wave $A\cos(2\pi ft)$. By how much is the phase of the resultant wave shifted in comparison with the original?

Answer: The resultant wave is $A\sin(2\pi ft)+\frac{1}{2}A\cos(2\pi ft)$,
i.e. $\sqrt{5/4}\,A\sin(2\pi ft+\phi)$ where $\phi=\tan^{-1}(\frac{1}{2})$.
Hence, $\phi=0.4637$ radians $=26.57°$.

EXERCISES 4D

1. Find the values of B and C when
   (a) $12\sin A+5\cos A$ is expressed in the form $B\sin(A+C)$;
   (b) $2\cos A+3\sin A$ is expressed in the form $B\cos(A+C)$.
2. The wave $10\sin(2\pi ft)$ is mixed with the wave $B\cos(2\pi ft)$. What is the value of B if the resultant wave has a phase shift of exactly $\pi/3$ with respect to the original sine wave?

5

Complex numbers

Complex numbers are numbers containing a multiple of the square root of -1. Because the product of two integer, rational or irrational numbers, either both positive or both negative, is positive, there is no real number which, multiplied by itself, will yield -1. However, before examining the kind of number which does yield -1 when squared, it is helpful to look at the arithmetic of surds.

5.1 QUADRATIC SURDS

A surd is the irrational root of a rational number, e.g. $\sqrt{2}$, $\sqrt[5]{95/7}$, $\sqrt[3]{3/2}$, etc. If a, b, and c are positive rational numbers, $\sqrt{a}$ is called a pure quadratic surd and $b \pm c\sqrt{a}$ is called a mixed quadratic surd. A quadratic surd is thus just a square root and is the only kind of surd we need consider for the present purposes.

A pocket calculator gives $\sqrt{5}=2.2360679\ldots$ but any such answer, to however many places of decimals, is an approximation since $\sqrt{5}$ is an irrational number and cannot be written as a finite string of digits. The calculator similarly provides an approximation to $3+\sqrt{5}$ and in the calculation of $y=(9-\sqrt{5})/(3+\sqrt{5})$ three approximations are made – one for the numerator, one for the denominator, and one for the calculation of the division. It is better practice to rearrange the fraction defining y in such a way as to reduce the approximations. This can be done by multiplying the top and bottom of the fraction by $(3-\sqrt{5})$. (The trick of multiplying through an expression by p/p, i.e. by 1, with a suitably chosen form of p, is often used in mathematics. We have met it before in Section 4.7 and will meet it several times again. It doesn't change the value of the expression, but simplifies it and so makes it more tractable.) The result of the multiplication is

$$y=\frac{(9-\sqrt{5})(3-\sqrt{5})}{(3+\sqrt{5})(3-\sqrt{5})}=\frac{27-12\sqrt{5}+(\sqrt{5})^2}{9-5}=\frac{32-12\sqrt{5}}{4},$$

$$=8-3\sqrt{5}, \tag{5.1}$$

so now y is calculated more easily than in its original form and with only one approximation.

In general, in manipulating quantities like $m+n\sqrt{a}$, where m, n, and a are rational numbers or integers (with a positive) and $\sqrt{a}$ is irrational, it is convenient to carry out the procedures of arithmetic but keeping the two halves of the expression, the rational and the irrational, separate except for acknowledging that the product of two irrationals is in this case rational since $(r\sqrt{a})(s\sqrt{a})=rsa$. This can be done explicitly by writing the mixed surd as a number pair (m, n) with rules of addition and multiplication, viz.

$$\left.\begin{array}{lll}\text{for addition,} & (m, n)+(u, v)=(m+u, n+v) \\ \text{for multiplication,} & (m, n)(u, v)=(mu+anv,\ mv+nu)\end{array}\right\} \tag{5.2}$$

Example 5(a) Putting $a=3$ in equation (5.2), compare the result of the multiplication $(4, 2)\cdot(3, -1)$ with the multiplication in the ordinary manner of $(4+2\sqrt{3})\cdot(3-\sqrt{3})$.

Answer: Using the multiplication rule of equation (5.2), the number pair product is $(4\cdot3+2\cdot(-1)\cdot3,\ 4\cdot(-1)+3\cdot2)$, i.e. $(6, 2)$. The multiplication of mixed surds gives

$$4\cdot3-4\cdot\sqrt{3}+6\cdot\sqrt{3}-2\cdot(\sqrt{3})^2=6+2\sqrt{3}.$$

The latter answer obviously corresponds exactly to the number pair answer.

Example 5(b) For number pairs obeying the rules given in equation (5.2), show that $(f, g)/(h, 0)=(f/h, g/h)$ and hence simplify $(f, g)/(j, k)$.

Answer: By the multiplication rule of equation (5.2)

$$\left(\frac{f}{h}, \frac{g}{h}\right)\cdot(h, 0)=\left(\frac{f}{h}\cdot h+a\cdot\frac{g}{h}\cdot0, \frac{f}{h}\cdot0+\frac{g}{h}\cdot h\right)=(f, g).$$

Hence,

$$\frac{(f, g)}{(h, 0)}=(f/h, g/h).$$

To deal with $(f, g)/(j, k)$ use the same device as in the simplification of y in equation (5.1) and multiply the fraction by $(j, -k)/(j, -k)$. Then the denominator becomes

$$(j, k)(j, -k)=(j^2-ak^2, 0)=(\alpha, 0)$$

if, for brevity, we write $\alpha=j^2-ak^2$. The numerator becomes $(f, g)(j, -k)=(fj-agk, gj-fk)$. Thus the fraction becomes

$$\left(\frac{fj-agk}{\alpha}, \frac{gj-fk}{\alpha}\right).$$

EXERCISES 5A

1. Using the rules of equation (5.2), with $a=2$, simplify $(1,0)/(1,1)$.
2. Use the binomial theorem to express $(m+n\sqrt{a})^r$ in the form of a number pair obeying the rules of equation (5.2) when r is an integer and confirm the validity of your answer for the case $m = n = 1$ and $r = a = 3$.

5.2 COMPLEX NUMBERS

I have elaborated the arithmetic of surds in order now to be able to emphasize the point that the arithmetic of complex numbers is exactly that given by equation (5.2) (and by Question 2 of exercises 2B) with a taking the value -1. And because complex numbers will play an important part in several topics of later chapters, the rules by which they are manipulated are here restated explicitly:

$$\left.\begin{array}{ll}\text{addition} & (m,n)+(u,v)=(m+u,n+v)\\ \text{multiplication} & (m,n)(u,v)=(mu-nv,mv+nu)\end{array}\right\} \qquad (5.3)$$

In dealing with surds, we used $\sqrt{a}$ where a is a rational and $\sqrt{a}$ an irrational number. In dealing with complex numbers, we are using $\sqrt{-1}$; -1 is an entirely acceptable rational number (which in this case, obviously, is a negative integer) but $\sqrt{-1}$ is a new concept which for historical reasons is called an imaginary number. Actually, it is *the* imaginary number because, since $\sqrt{pq}=\sqrt{p}\sqrt{q}$, we can always write $\sqrt{-p}$ as $\sqrt{p}\sqrt{-1}$. Thus $\sqrt{-1}$ is the only imaginary number used. Exactly corresponding to a mixed quadratic surd, the complex number is written as $a+b\sqrt{-1}$ or as (a, b) in the number pair notation; in either notation, a is called the real part and b the imaginary part of the number.

It is often necessary to symbolize the quantity $\sqrt{-1}$ itself. Mathematicians, physicists and many other scientists use i for this purpose, but electrical and electronic engineers use j because they claim they need to let i stand for electric current, which physicists usually indicate by I. Henceforth in this book I shall use i to mean $\sqrt{-1}$ because I was brought up that way. Thus a complex number z might be written as (x, y) or as $z = x + \mathrm{i}y$.

5.3 THE ARGAND DIAGRAM

The complex number z, symbolized by the number pair (x, y), is conveniently represented as a point P in a plane governed by rectangular axes as shown in Fig. 5.1. Of the two axes, the horizontal one in Fig. 5.1 is called the real axis (indicated by $\mathscr{R}$) and the other is called the imaginary axis (indicated by $\mathscr{I}$). Then P is plotted with x as the coordinate along the $\mathscr{R}$-axis and y as the coordinate along the $\mathscr{I}$-axis.

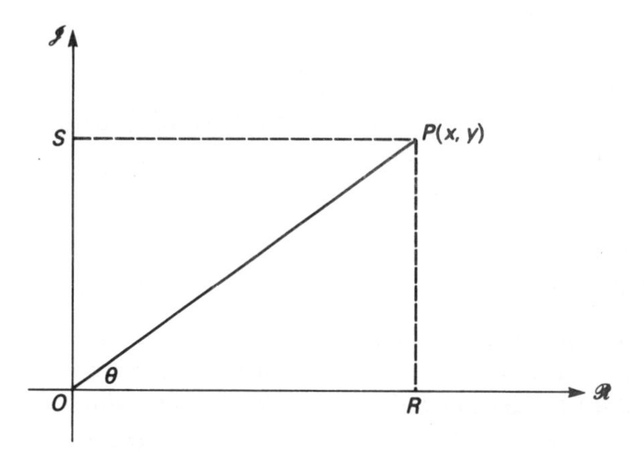

Figure 5.1

This representation is called the Argand diagram, but note that it provides a strictly diagrammatic representation of complex numbers. The point P is not the number z in the same sense that the point O in Fig. 5.1 is the intersection of the real and imaginary axes. The axes are entities in the diagram and they intersect at an actual point which can be given a name (O). The point P is an actual point, but it isn't an actual number.

The number z, indeed, can in some ways be considered as being like the displacement needed to get from O to P or as being like any equal displacement one might make in the plane in the same direction as $O \rightarrow P$. You will get a better appreciation of this from consideration of the diagrammatic representation of the addition of two complex numbers. First, however, there are some other matters to note about the representation shown in Fig. 5.1.

To begin with, if R and S are the feet of the perpendiculars dropped from P on to the real and imaginary axes, it is evident that x is represented by the length OR (=SP) and y by the length OS ($=RP$). We now have that $x = OP \cos \angle POR$ and $y = OP \sin \angle POR$. Write r for the length OP and θ for the angle $\angle POR$. The $x = r \cos \theta$ and $y = r \sin \theta$. But also, by Pythagoras, $OP^2 = OR^2 + RP^2 = x^2 + y^2$, or, looked at the other way round,

$$\begin{aligned} x^2 + y^2 &= r^2\cos^2\theta + r^2\sin^2\theta \\ &= r^2(\cos^2\theta + \sin^2\theta) \\ &= r^2 \end{aligned}$$

since $\sin^2\theta + \cos^2\theta = 1$, as shown in Table 4.3. Thus if $z = x + \mathrm{i}y$, it follows that we could also write

$$\begin{aligned} z &= r \cos \theta + \mathrm{i}\, r \sin \theta \\ &= r(\cos \theta + \mathrm{i} \sin \theta). \end{aligned}$$

Sometimes it is convenient to write $x=\mathscr{R}(z)$ and $y=\mathscr{I}(z)$, indicating that the mapping operation of the Argand diagram involves a function $\mathscr{R}$ which can operate on z to extract its real part and another function $\mathscr{I}$ to extract its imaginary part.

Finally, in this part of the exposition, here are some important and frequently used words. The length OP, i.e. r, is called the **modulus** or magnitude of z and θ is called the **argument** or amplitude of z. The symbol for modulus is two vertical bars enclosing the number in question, so write $r=|z|$. The modulus or magnitude function always returns a positive quantity; that is, where $z=x+iy$, $|z|$ is the positive square root of x^2+y^2. Indeed, this understanding is extended to real numbers so that, as an example, $-k$ might be thought of as the real part of the complex number $-k+i0$ (i.e. a complex number with a zero imaginary part). Its modulus is then $\sqrt{(-k)^2+0^2}$ which is k (i.e. $+k$ and not $-k$). The argument of z, namely the angle θ, (sometimes written as arg(z) or am(z)) is specified by $\theta=\tan^{-1}(y/x)$, but with the restriction that $-\pi<\theta<\pi$. With this restriction, θ is said to take on the principal value of the argument.

5.4 ADDITION OF COMPLEX NUMBERS

Suppose there are two complex numbers $z_1=x_1+iy_1$ and $z_2=x_2+iy_2$ and represented by P and Q at (x_1, y_1) and (x_2, y_2), respectively, in the Argand diagram as shown in Fig. 5.2. The rules for adding complex numbers are given by equation (5.3). But those rules are for numbers; what about points? There is no meaning to the idea of adding points in the plane, but one *can* add displacements. In Fig. 5.2 the appropriate displacements are indicated by the various full and dashed lines. It can be seen that the displacement $O\to P$ followed by the displacement $P\to S$ (which is parallel to and hence

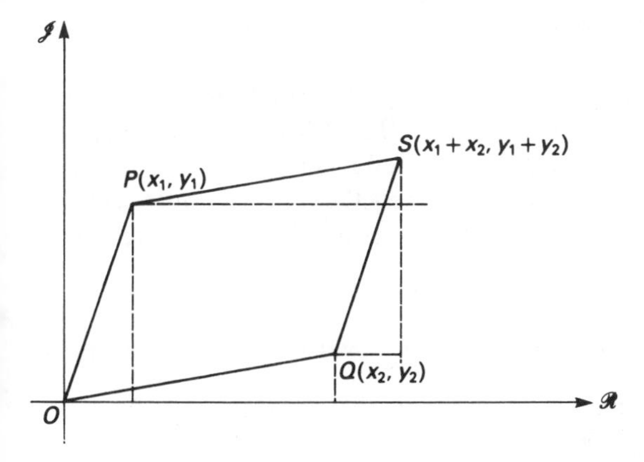

Figure 5.2

equivalent to the displacement $O \to Q$) represents the addition of z_2 to z_1. Alternatively, the displacement $O \to Q$ followed by the displacement $Q \to S$ represents adding z_1 to z_2. Both manoeuvres end at S as they must since S in the diagram represents z_1+z_2 which in turn, by the laws of arithmetic of Chapter 2, equals z_2+z_1.

Thus in the Argand diagram, the addition of two complex numbers is achieved geometrically by completing a parallelogram, the figure $OPSQ$ in Fig. 5.2, of which two adjacent sides (OP and OQ) represent the numbers.

5.5 MULTIPLICATION OF COMPLEX NUMBERS

The rule of multiplying together the two complex numbers z_1 and z_2 is given by equation (5.3). Thus $z_1 z_2=(x_1 x_2-y_1 y_2, x_1 y_2+x_2 y_1)$. If now the modulus and argument of z_1 are r_1 and θ and of z_2 are r_2 and ϕ, we have that $x_1=r_1 \cos \theta$, $y_1=r_1 \sin \theta$, etc. Substituting these values into the expression for the product, we have

$$\begin{aligned} z_1 z_2 &= (r_1 \cos \theta r_2 \cos\phi - r_1 \sin \theta r_2 \sin\phi,\ r_1 \cos \theta r_2 \sin\phi + r_1 \sin \theta r_2 \cos\phi) \\ &= (r_1 r_2 \cos(\theta+\phi),\ r_1 r_2 \sin(\theta+\phi)) \qquad \text{from Table 4.3.} \end{aligned}$$

If we put w as the product of z_1 and z_2 (i.e. put $w=z_1 z_2$) and say that the modulus of w is ρ and its argument is ψ, then we would have that $w=\rho(\cos \psi + \mathrm{i} \sin \psi)$ or that w was represented on the Argand diagram by the point $(\rho \cos \psi, \ \rho \sin \psi)$. Comparing this with the value of $z_1 z_2$ just calculated, we reach an important result: if two complex numbers are multiplied together, the modulus of the product is the product of the moduli of the two numbers and the argument of the product is the sum of the arguments of the two numbers (or else differs from that sum by 2π in order to abide by the restriction of lying between $-\pi$ and π). There are corresponding statements which apply to the quotient of two numbers; these involve the quotient of the moduli of the numbers and the difference of their arguments.

There is an elaborate way of constructing this result in the Argand diagram, but it is hardly ever used. However, Fig. 5.3 illustrates the result. If the numbers represented by the points P (modulus r_1, argument θ) and Q (modulus r_2, argument ϕ) are multiplied together, the product can be represented by the point R where OR is at an angle $\theta+\phi$ to the real axis and is of length $r_1 r_2$.

Example 5(*c*) Given that $z_1=4+\mathrm{i}2$ and $z_2=3-\mathrm{i}$, calculate z_1+z_2 and $z_1 z_2$.

Answer: This is simply a matter of applying the rules of equation (5.3), giving $z_1+z_2=7+\mathrm{i}$ and $z_1 z_2=14+\mathrm{i}2$.

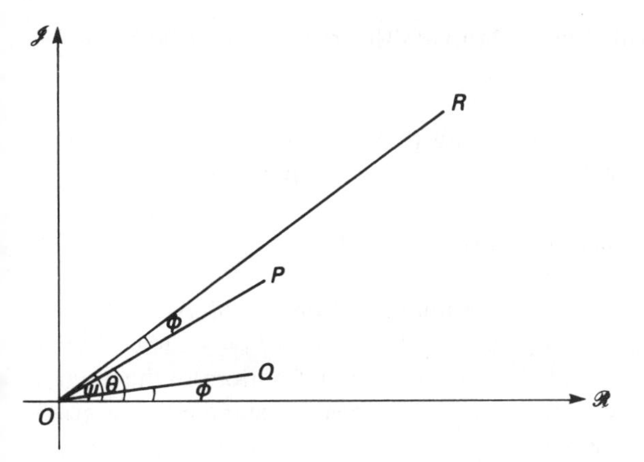

Figure 5.3

Example 5(*d*) Calculate $(2+i3)^3$.

Answer: One could do this by applying the rules of equation (5.3) twice, first to find $(2+i3)^2$ and then to find $(2+i3)((2+i3)^2)$. Alternatively, one could use a binomial expansion:

$$\begin{aligned}(2+i3)^3 &= 2^3+3.2^2\cdot(i3)+3.2\cdot(i3)^2+(i3)^3\\ &= 8+36i+54i^2+27i^3.\end{aligned}$$

But $i^2=-1$ and $i^3=-i$. Hence $(2+i3)^3=-46+i9$.

EXERCISES 5B

1. If $z_1=p+i2q$ and $z_2=2p-iq$, calculate z_1+z_2 and z_1z_2.
2. Show that $(\cos\pi/3+i\sin\pi/3)^3=-1$, i.e. show that $(\cos\pi/3+i\sin\pi/3)$ is a cube root of -1.

5.6 DE MOIVRE'S THEOREM

If z_1 and z_2 are two complex numbers of modulus 1 and with arguments θ and ϕ, respectively, then $z_1=\cos\theta+i\sin\theta$, $z_2=\cos\phi+i\sin\phi$, and $z_1z_2=\cos(\theta+\phi)+i\sin(\theta+\phi)$, according to the result of the last section. If now we let $z_1=z_2$ so that $\phi=\theta$, then

$$z_1^2=(\cos\theta+i\sin\theta)^2=\cos 2\theta+i\sin 2\theta.$$

Continuing to multiply by z_1 we obtain de Moivre's theorem:

$$(\cos\theta+i\sin\theta)^n=\cos n\theta+i\sin n\theta. \tag{5.4}$$

It can be shown very easily that de Moivre's theorem holds for both positive and negative values of n.

Example 5(e) If $z=\cos\theta+\mathrm{i}\sin\theta$, show that $z^{-1}=\cos\theta-\mathrm{i}\sin\theta$ and that $z^{-3}=\cos 3\theta-\mathrm{i}\sin 3\theta$. Show that $(z+z^{-1})^3=8\cos^3\theta$ and hence that

$$\cos 3\theta=4\cos^3\theta-3\cos\theta.$$

Answer: The formula for z^{-1} may be obtained from de Moivre's theorem by putting $n=-1$ and using $\cos(-\theta)=\cos\theta$ and $\sin(-\theta)=-\sin\theta$. Alternatively, since $z^{-1}=(\cos\theta+\mathrm{i}\sin\theta)^{-1}$ multiply top and bottom of the fraction by $\cos\theta-\mathrm{i}\sin\theta$ and since $\cos^2\theta-\mathrm{i}^2\sin^2\theta=\cos^2\theta+\sin^2\theta=1$, the result follows. Similarly for z^{-3}. Now,

$$\begin{aligned} z+z^{-1} &= (\cos\theta+\mathrm{i}\sin\theta)+(\cos\theta-\mathrm{i}\sin\theta) \\ &= 2\cos\theta. \end{aligned}$$

Therefore $(z+z^{-1})^3=(2\cos\theta)^3=8\cos^3\theta$. But also

$$\begin{aligned} (z+z^{-1})^3 &= z^3+3z^2(z^{-1})+3z(z^{-1})^2+(z^{-1})^3 \\ &= z^3+3(z+z^{-1})+z^{-3}. \end{aligned}$$

Hence, using de Moivre and the results just obtained,

$$\begin{aligned} 8\cos^3\theta &= \cos 3\theta+\mathrm{i}\sin 3\theta+6\cos\theta+\cos 3\theta-\mathrm{i}\sin 3\theta \\ &= 2\cos 3\theta+6\cos\theta. \end{aligned}$$

Divide by 2 and rearrange to give $\cos 3\theta=4\cos^3\theta-3\cos\theta$.

EXERCISES 5C

1. By examining $(z-z^{-1})^3$ where $z=\cos\theta+\mathrm{i}\sin\theta$, show that $\sin 3\theta=3\sin\theta-4\sin^3\theta$.
2. Use de Moivre's theorem to find the square root of i. (Hint: look at $\cos\pi/2+\mathrm{i}\sin\pi/2$.)

5.7 THE COMPLEX CONJUGATE

On a number of occasions in the past few pages a complex number has been multiplied by another in such a way as to give a result with no imaginary part in the product. The pattern of these multiplications is always the same: if one number is $x+\mathrm{i}y$ it is multiplied by $x-\mathrm{i}y$. The reason is that the product $(a+b)(a-b)=a^2-b^2$ whatever the values of a and b; there are no cross-products, i.e. no terms in ab. In the case of $(x+iy)(x-iy)$ the result, $x^2-(iy)^2=x^2+y^2$, is entirely real and is always positive since x and y are both real numbers, the square of a real number is necessarily positive, and the sum of two positive numbers must be positive.

The two numbers $x+iy$ and $x-iy$ are called **complex conjugates** of each other. If z is one of these numbers, the other is symbolized by z^*; that is, the superscript asterisk implies the original number but with the sign of the imaginary part inverted – positive-to-negative or negative-to-positive as the case may be. The product of the complex conjugates is not only real, but its magnitude is the square of the modulus of z and the modulus of z^*. Alternatively expressed, $\sqrt{zz^*}=|z|=|z^*|$.

The complex conjugate is particularly useful in simplifying complex fractions by clearing imaginary parts from the denominator, and also in determining the modulus of a number. A useful understanding of the relationship between a number and its complex conjugate is given by the Argand diagram. If, as in Fig. 5.4, a complex number $p+iq$ is plotted at the point P in the Argand diagram, its complex conjugate, $p-iq$, will be at the point P' which is the reflection of P in the real axis. If the number represented by the point P has the argument α, then its complex conjugate represented by P' has the argument $-\alpha$, as shown in Fig. 5.4. It is then obvious that the sum of the two numbers, obtained in the Argand diagram by completing the parallelogram containing O, P, and P', will lie on the real axis. But equally, the product of the two numbers, with argument the sum of their arguments, will also lie on the real axis, since its argument will be zero.

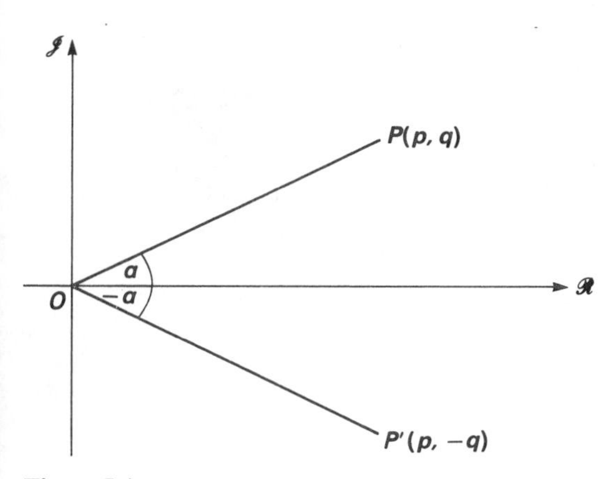

Figure 5.4

Example 5(f) Find the modulus of $(2+i4)/(1+i)$.

Answer: First simplify the fraction; multiply numerator and denominator by the complex conjugate of the denominator, i.e.

$$\frac{2+i4}{1+i}=\frac{(2+i4)(1-i)}{(1+i)(1-i)}=\frac{2+4-i2+i4}{2}$$

$$=\frac{6+i2}{2}=3+i.$$

The modulus is given by $\sqrt{(3+i)(3-i)}=\sqrt{9+1}=\sqrt{10}.$

EXERCISES 5D

1. Find the modulus of $(4+i7)/(3-i)$.
2. Express in the form $x+iy$ the expression

$$(\cos 3\theta+i\sin 3\theta)^2/(\cos\theta-i\sin\theta)^5.$$

6

Graphs

You will quite certainly have had considerable experience of using graphs as illustrations of the relations between variables in many branches of science and also as illustrations of experimental results you yourself have obtained in laboratory experiments. This chapter is therefore quite short, merely recapitulating and drawing together many things you already know.

Note to begin with that if we graph a functional relationship $y=f(x)$ we are doing something distinctly different from effecting a mapping from the domain of x into the domain of y, as discussed in Chapter 3. The plane of Fig. 6.1 is, in a sense, the domain of both x and y and a graph in it is a

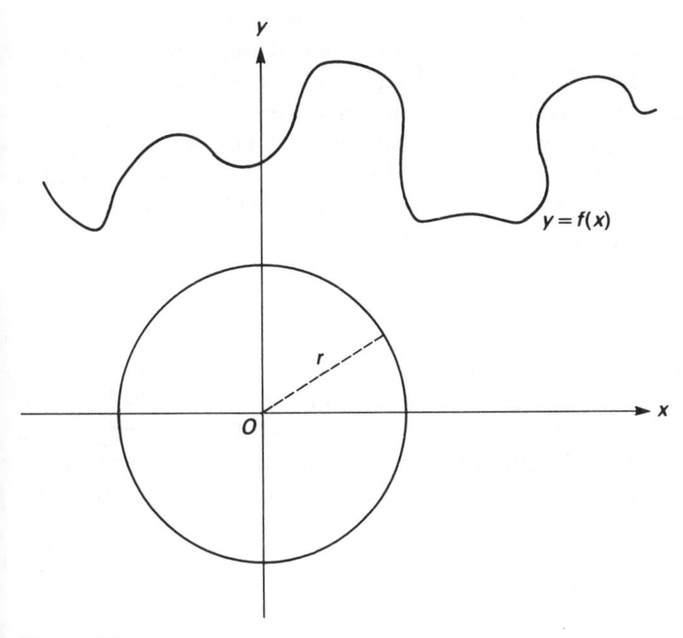

Figure 6.1

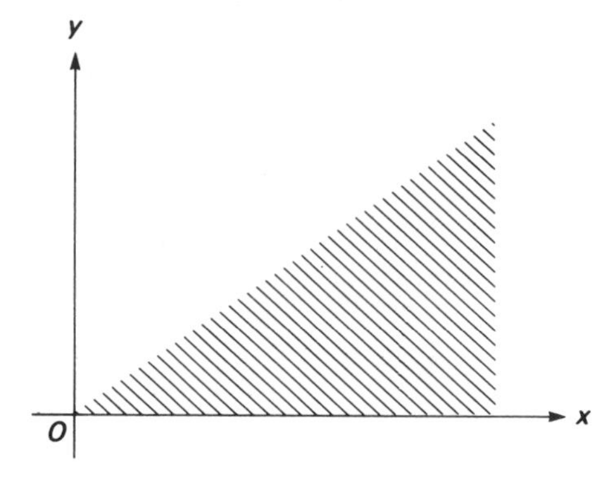

Figure 6.2

joining up of the positions of all the points (x, y) which satisfy the specified functional relationship. The relationship may be explicit, of the form $y=f(x)$, i.e. with y alone on the l.h.s. and x the only variable on the r.h.s.; or it may be implicit, for example as given by the equation $x^2+y^2-r^2=0$ which graphs the circle shown in Fig. 6.1 and to which we return below. Sometimes an implicit relationship can be manipulated into an explicit one (e.g. in the case of the circle, $y=\pm\sqrt{r^2-x^2}$), but sometimes it can't be.

In mathematicians' terminology, the representation of all the positions which satisfy a given relationship is called a **locus** (plural: loci). A locus is not necessarily a line in the plane as in Fig. 6.1. It might be an area; for example, $0<y<x$ specifies the shaded region in Fig. 6.2. And in a three-dimensional graph, a locus might be a curved surface or a volume. It is, of course, assumed here that x and y are real numbers. The planes of Figs 6.1 and 6.2 are not to be confused with the complex plane of the Argand diagram used in Chapter 5, in which the ordinate axis placed the imaginary component of a complex number.

6.1 THE STRAIGHT LINE

A linear relationship between x and y, i.e. one of the form $rx+sy+t=0$ where x and y are variables and r, s, and t are constants, is so called because its locus is a straight line. Conversely, a relationship between x and y which involves either or both of them in powers other than 1 yields a locus which is a curved line. The equation for a straight line is often written in the form

$$y=mx+c. \tag{6.1}$$

Since $rx+sy+t=0$ is the same as

$$y=\left(-\frac{r}{s}\right)x-\frac{t}{s},$$

the two equations for a line are the same if we write m for $-r/s$ and c for $-t/s$. The form of equation (6.1) makes y clearly the dependent variable and x the independent one. Figure 6.3 shows a line AP inclined at an angle θ to the x-axis and with a construction of perpendicular lines, etc. where right angles are marked ⌐. For the point P with coordinates (x, y), it is obvious that $x = OC = AB$ and $y = PC = PB + BC$. But $PB/AB = \tan\theta$ so that $PB = AB\tan\theta = x\tan\theta$; and $BC = OA$. Thus $y = (\tan\theta)\cdot x + OA$. Hence in the equation $y = mx + c$, we can identify m with $\tan\theta$ and c with OA. The quantity $\tan\theta$ (i.e. the tangent of the angle made by the line with the x-axis) is called the gradient of the line and the quantity c is the intercept on the y-axis. If $c = 0$, the line passes through the origin of the axes, the point O, and if c is negative, the line intersects the y-axis below O. The angle θ is always taken to be in the range 0 to π. Thus if m is positive, θ lies between 0 and $\pi/2$ and if m is negative θ lies between $\pi/2$ and π.

In equation (6.1) the quantities m and c were implicitly considered as constants, but in general they are called parameters. The meaning of this often misused word is that the quantity to which it is applied is constant in a particular situation, but can vary from one situation to another. For any one straight line, m and c will have definite values, while x and y can have any values (though not independently). For another line, m and c will have different values while x and y remain the true variables. The parameters control the slope and position of the line; the variables control the position of points on the line.

As an example, consider the problem of finding the equation of the straight line which passes through the two points P at (X_1, Y_1) and Q at (X_2, Y_2) as shown in Fig. 6.4. From the construction shown in that figure,

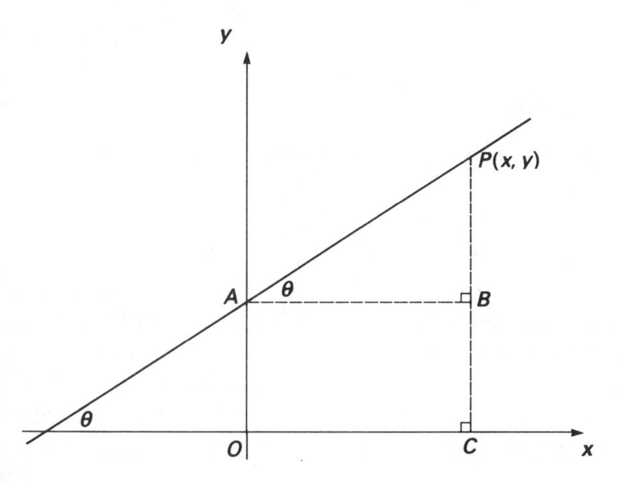

Figure 6.3

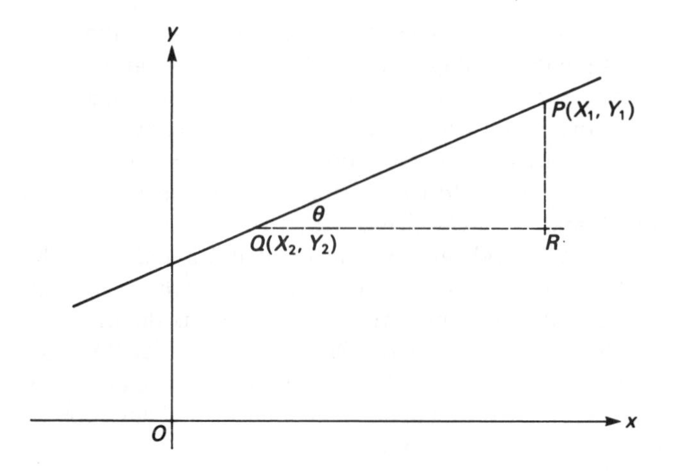

Figure 6.4

it is seen that

$$\tan\theta = \frac{PR}{RQ} = \frac{Y_1 - Y_2}{X_1 - X_2}.$$

But if the equation being sought is written in the form $y = mx + c$ we know that $m = \tan\theta$, so the value of m is now established. To find the value of c, the argument is that the line certainly passes through P, so the coordinate values X_1 and Y_1 must satisfy the equation $y = mx + c$. Hence

$$Y_1 = \frac{Y_1 - Y_2}{X_1 - X_2} X_1 + c$$

so that

$$y = \frac{Y_1 - Y_2}{X_1 - X_2} x + \left(Y_1 - \frac{Y_1 - Y_2}{X_1 - X_2} X_1 \right).$$

After some rearrangement, the line is given by the equation

$$(y - Y_1) = \left(\frac{Y_1 - Y_2}{X_1 - X_2} \right)(x - X_1).$$

Suppose there is a line L_1, say, inclined to the x-axis at an angle θ, so that its gradient, m_1, has the value $\tan\theta$. If another line, L_2, lies perpendicular to L_1, then L_2 will be inclined at an angle $\theta + \pi/2$ to the x-axis. In that case, m', the gradient of L_2, will have the value $\tan(\theta + \pi/2)$. But $\tan(\theta + \pi/2) = -\cot\theta = -1/\tan\theta$. Hence we can conclude that if two straight lines with gradients m and m' are perpendicular to each other, $mm' = -1$, and vice versa.

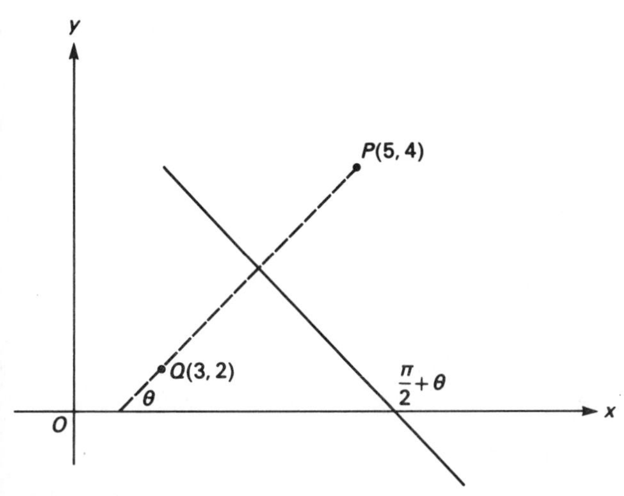

Figure 6.5

Example 6(a) Find the equation of the line which is the perpendicular bisector of the line PQ where P is the point (5, 4) and Q is the point (3, 2).

Answer: If the line PQ is inclined at an angle θ to the x-axis, its gradient is

$$m=\frac{4-2}{5-3}=1.$$

The line perpendicular to this one therefore has a gradient $m'=-1$ as in Fig. 6.5. The mid-point of PQ is given by the coordinates $((5+3)/2, (4+2)/2)$, i.e. is the point (4, 3). This point lies on the line so if the line is $y=m'x+c$ and $m'=-1$, then $3=-1.4+c$ which gives $c=7$ and the equation of the line is $y=-x+7$. (It would be neater to write the equation as $x+y-7=0$.)

EXERCISES 6A

1. What is the equation of the line which passes through the origin, parallel to the line through the points P and Q of example 6(a)?
2. If a line passes through the point (2, 8) and is inclined at an angle of $\pi/4$ to the x-axis, where does it intersect the y-axis?

6.2 THE CIRCLE

A circle drawn with its centre at the origin O and with radius a is the locus of all points like P at (x, y) in Fig. 6.6 such that $OP=a$. For the point P we

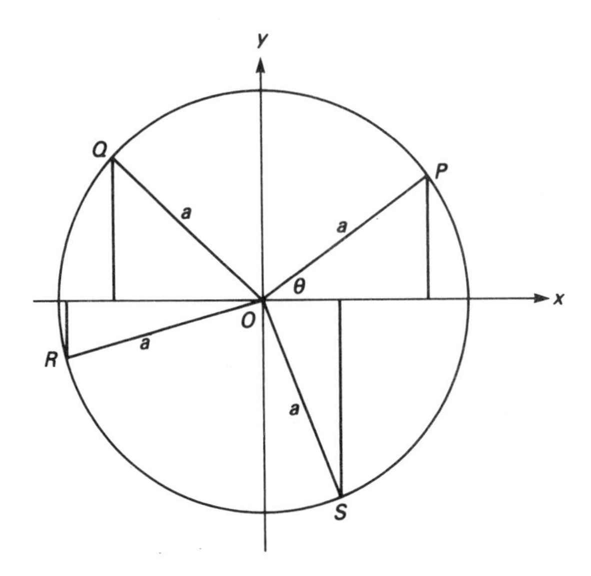

Figure 6.6

have

$$x^2+y^2=a^2 \tag{6.2}$$

and this is true of all other points on the circle, like Q, R, and S in Fig. 6.6. Hence equation (6.2) is the equation for the locus of all points at a distance a from O, i.e. is the equation of the circle.

If in Fig. 6.6 the angle POx is called θ, then $x=a\cos\theta$ and $y=a\sin\theta$. In this case $x^2+y^2=a^2(\cos^2\theta+\sin^2\theta)=a^2$, so there is no departure from the statement given in equation (6.2). The expression of x and y in terms of a third variable, here the angle θ, is called a parametric representation – in this case, a parametric representation of the equation for a circle. The name comes about because a particular value assigned to the parameter θ determines the values of the variables x and y. The parametric form of the equation of a circle appears quite frequently in various applications of physics and electronics and thereby in such topics as biological signal analysis. In these applications θ is typically a function of time, t, or distance d. For example, if $\theta=2\pi\nu t$, y changes sinusoidally as t increases steadily and so does x, but $\pi/2$ out of phase with y. (Remember that $\sin(\pi/2+\theta)=\cos\theta$, as shown in Table 4.3.) In these circumstances, x and y have values between $-a$ and $+a$ and oscillate between the extremes with frequency ν; i.e. if x and y have particular values at some moment when $t=t_0$, then t must increase by an amount ν^{-1} before x and y return to those values. This comes about because $2\pi\nu(t_0+\nu^{-1})=2\pi\nu t_0+2\pi$, and $\cos(2\pi+\theta)=\cos\theta$, with a

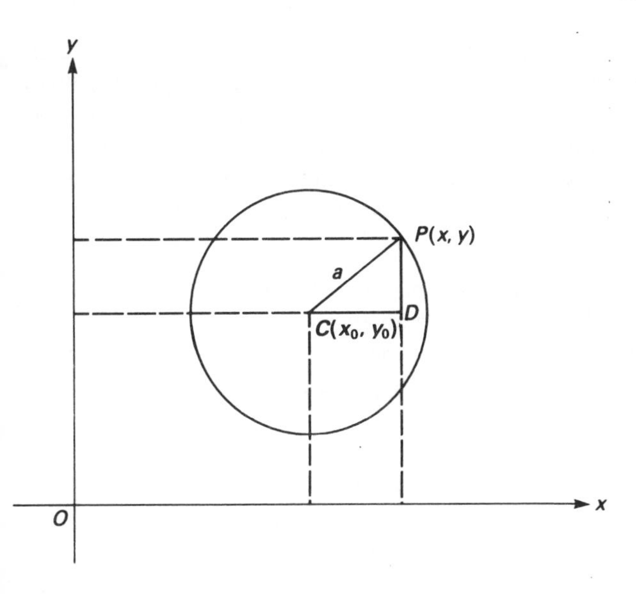

Figure 6.7

corresponding result for the sine function. Alternatively, if we were to concentrate on the point P, as time progressed steadily we would see P move steadily round the circle.

If the circle is to be drawn about a centre at C, the point with coordinates (x_0, y_0), then $CP = a$. Applying Pythagoras' theorem to the triangle PCD in Fig. 6.7, $CD^2 + DP^2 = CP^2$. Hence the equation for the circle becomes:

$$(x - x_0)^2 + (y - y_0)^2 = a^2. \tag{6.3}$$

This equation becomes equation (6.2) if C is at O and $x_0 = y_0 = 0$.

Note that while equation (6.3) provides the locus of the circle drawn in Fig. 6.7, the inequality $(x - x_0)^2 + (y - y_0)^2 \leqslant a^2$ is a specification of all the points inside the circle or on its circumference.

6.3 CONIC SECTIONS

Writing out equation (6.3) in full gives

$$x^2 + y^2 - 2x_0x - 2y_0y + (x_0^2 + y_0^2 - a^2) = 0.$$

This is a particular case of a more general equation

$$ax^2 + 2hxy + by^2 + 2fx + 2gy + c = 0. \tag{6.4}$$

Depending on the values taken by the parameters a, b, c, f, g, and h, the loci defined by equation (6.4) are called the conic sections. These are geometric figures which have been studied since the time of the Greek mathematicians working more than 2000 years ago. If a double cone is imagined, i.e. two identical cones, one mounted upside-down on top of the other, a plane slicing through the double cone will reveal a shape in the cut surface as in Fig. 6.8. Suppose that the sides of the cone make an angle α with the axis of the cone and let a plane cutting the cone make an angle β with the axis. Then:

1. the plane A, with $\beta=\pi/2$, reveals a circle in the section;
2. the plane B, with $\pi/2>\beta>\alpha$, reveals an ellipse;
3. the plane C, with $\beta=\alpha$, reveals a parabola;
4. the plane D, with $0<\beta<\alpha$, reveals a hyperbola;

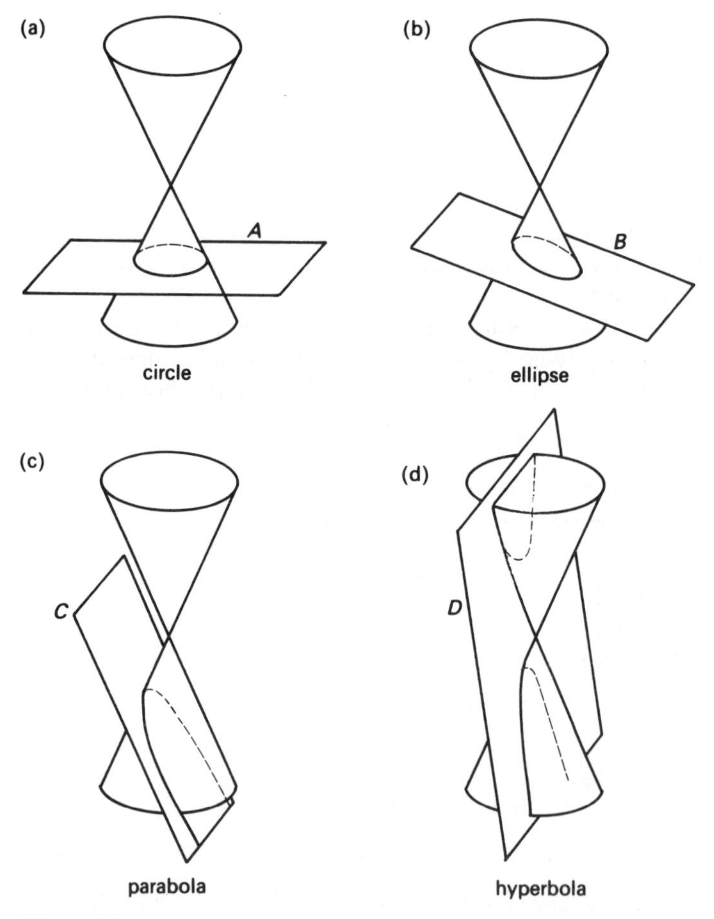

Figure 6.8

5. a plane through the apex P of the cone, and with $0<\beta<\alpha$, reveals a pair of intersecting straight lines.

The general equation for the conic sections, equation (6.4), provides the locus of a circle if $a=b$ and also $h=0$, as can be inferred from equation (6.3). In the more general cases, if $h^2<ab$ the locus is an ellipse, if $h^2=ab$ it is a parabola, and if $h^2>ab$ it is a hyperbola. The condition for equation (6.4) to provide the locus of a pair of straight lines is the rather elaborate relation $abc+2fgh-af^2-bg^2-ch^2=0$.

The equation for a circle centred at the origin, equation (6.2), can be written as

$$\frac{x^2}{a^2}+\frac{y^2}{a^2}=1.$$

The equation for an ellipse centred at the origin and with its axes along the coordinate axes is given by the similar equation

$$\frac{x^2}{a^2}+\frac{y^2}{b^2}=1.$$

In this case (as shown in Fig. 6.9) the ellipse stretches from $-a$ to $+a$ along the x-axis and from $-b$ to $+b$ along the y-axis. The parametric form of this equation of the ellipse is $x=a\cos\theta$ and $y=b\sin\theta$. As with the corresponding equation for a circle, θ might be a function of time and an account involving such a description is sometimes given for elliptically polarized light, encountered in studies of circular dichroism and optical rotatory dispersion of biochemical compounds. For ellipses less constrained in position and orientation, equation (6.4) with the appropriate relation between a, b, and h is required, but such figures appear mostly in astronomy and rarely in laboratory-based sciences.

Equally, in laboratory-based sciences, parabolas are most often found in

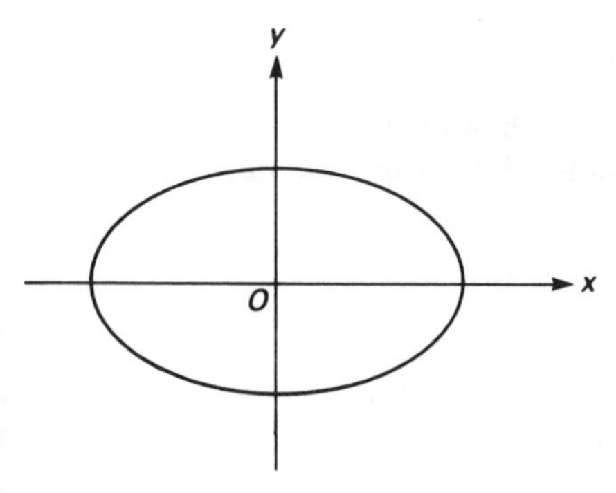

Figure 6.9

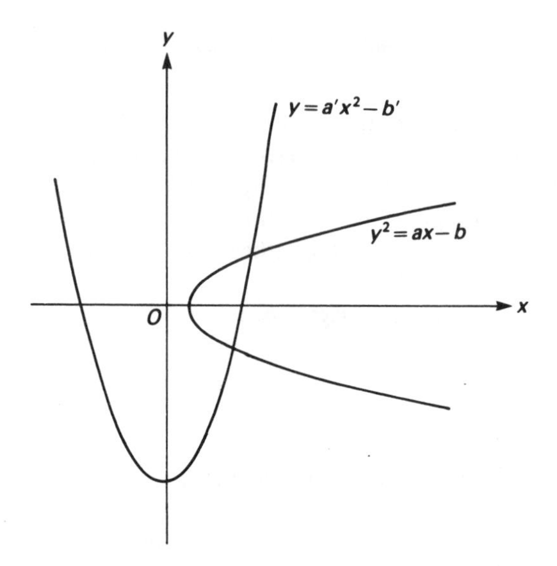

Figure 6.10

graphs of the form $y=ax^2+b$ or else $y^2=cx+d$ as shown in Fig. 6.10. In these circumstances hyperbolae too are found only in limited situations. A hyperbola is a figure which, far from its centre, approximates to a pair of straight lines, its asymptotes, as shown in Fig. 6.11. But, in the graphing of experimental results, the straight lines are usually the coordinate axes at right angles to each other, in which case the conic section is called a rectangular hyperbola as shown in Fig. 6.12. The appropriate formula is $xy=k$, where k is a constant, but the asymptotes are sometimes found to be parallel to, though displaced from, one or other of the axes. The equation to be graphed or to be fitted to experimental results may then appear as something like

$$x=\frac{ky}{y+l}+m$$

where l and m are other constants (although m might be zero).

When experimental results or a scientific law or hypothesis yield a graph

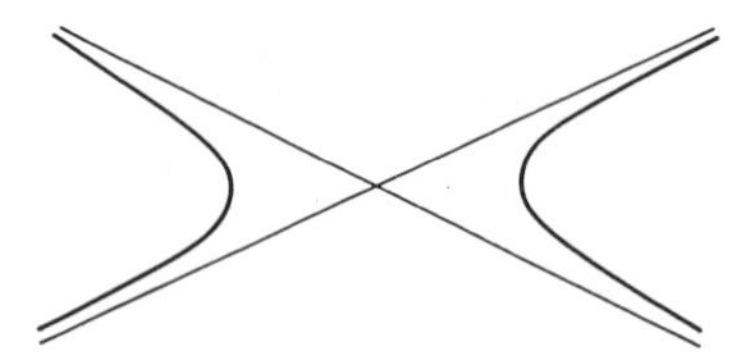

Figure 6.11

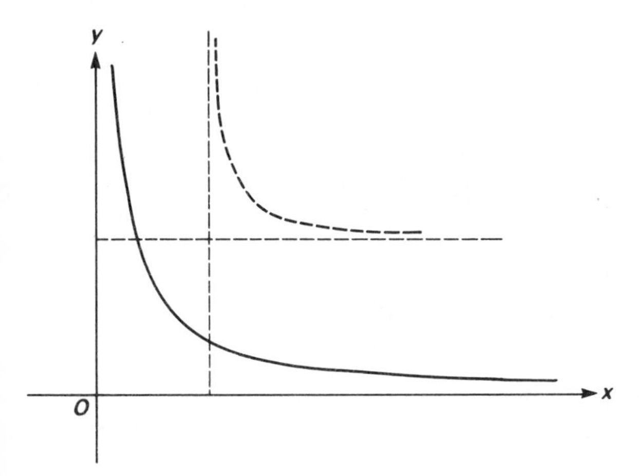

Figure 6.12

which is a conic section, it is sometimes possible to use the well-known geometry of the figures to make an important inference about the experimental system being studied. It is for this reason that I have listed some simple graphical features of the conic sections. However, the number of geometrical results is vast and if the matter arises in a practical way you must refer to a specialized text.

6.4 HIGHER-DEGREE PLANE CURVES

A linear equation in x and y provides a locus which is a straight line on a graph. Any other line not parallel to it will cut it in just one point. A quadratic equation, that is an equation with terms in x^2, y^2 and xy, as well as terms in x and y, with equation (6.4) as the most general form of such an equation, provides loci which are the conic sections. A straight line, if placed on the plane in such a position as to intersect one of these sections, will do so in two points as shown in Fig. 6.13. Similarly, a cubic equation, one with terms in x^3, x^2y, xy^2, and y^3, as well as terms like those in quadratic equations, can be intersected in up to three points by a straight line. In general, an equation with terms in x and y up to the power of n can be cut by a straight line in up to n points; but the details of the result depend in a complicated way on the geometry of the locus.

6.5 COORDINATE SYSTEMS

I have hitherto in this chapter dealt with points on graphs mostly in terms of two coordinates x and y which in turn are referred to two axes Ox (the

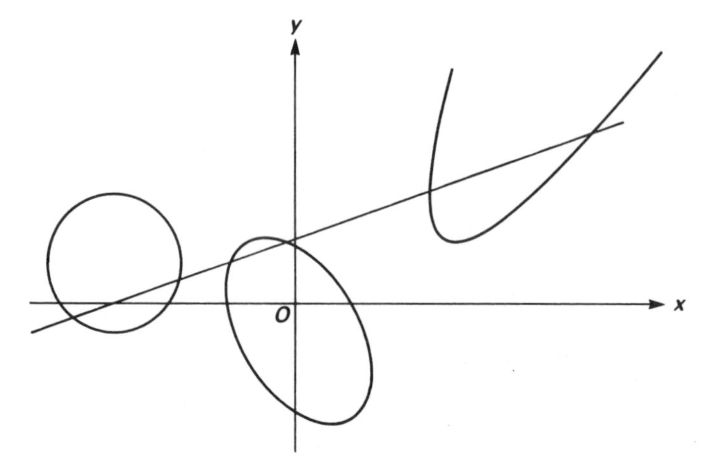

Figure 6.13

abscissa) and Oy (the ordinate) at right angles to each other. This system of locating points is called the orthogonal Cartesian system, 'orthogonal' implying that the axes are indeed at right angles, and 'Cartesian' honouring Descartes who introduced and popularized the method. However, any other unambiguous system of locating a point in a plane could be used. For example, the axes need not be orthogonal (at right angles to each other) and non-orthogonal Cartesian systems find application by crystallographers when they deal with rhombohedral, monoclinic, or triclinic crystals (Fig. 6.14).

However, the most important coordinate system after the orthogonal Cartesian one is the system of polar coordinates. This requires the specification only of an origin or pole, O, and an initial direction (Fig. 6.15). The

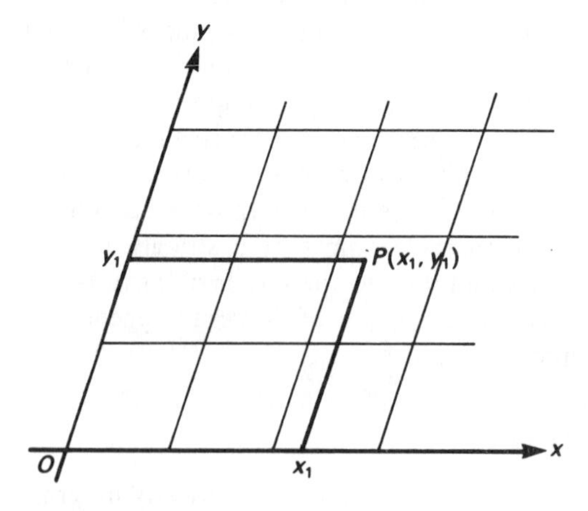

Figure 6.14

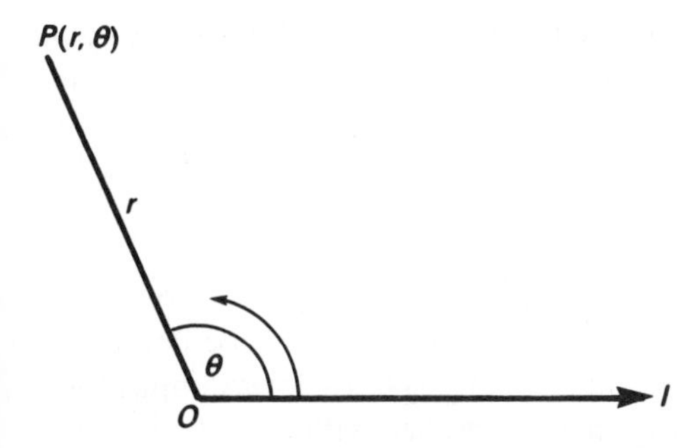

Figure 6.15

position of a point P is then specified by the angle POI $(=\theta)$ and the radial distance OP $(=r)$. Occasionally the use of polar coordinates (r, θ) is of advantage over Cartesian coordinates, but not often. A circle of radius a and centred at O is very easily described in polar coordinates, namely by the equation $r=a$; this is clearly simpler than the Cartesian description of the same circle. But a circle of radius a centred elsewhere in the plane, which in Cartesian coordinates is given by equation (6.3), becomes in polar coordinates $r^2-2Rr\cos(\theta-\alpha)+R^2=a^2$ if the centre is at a point C with polar coordinates (R, α).

In practice, the system of coordinates chosen to deal with a problem is determined in a fairly obvious way by the conditions of the problem. Thus, for example, a computerized tomographic (CT) scan involving rotation of the imaging system about a centre might be best dealt with by referring to polar coordinates, whereas a gamma camera making a linear scan over a patient would be described by Cartesian coordinates.

6.6 MULTIDIMENSIONAL SYSTEMS

The graphing of points can be extended from the plane of the paper into the air above the paper. That would involve adding to the x and y axes another axis z, say, and giving to each point the three coordinates (x, y, z). In the two-dimensional system, $x^2/a^2+y^2/a^2=1$ gave the locus of a circle and $x^2/a^2+y^2/b^2=1$ gave the locus of an ellipse. In the three-dimensional system,

$$\frac{x^2}{a^2}+\frac{y^2}{a^2}+\frac{z^2}{a^2}=1$$

gives a locus which is the surface of a sphere of radius a, and

$$\frac{x^2}{a^2}+\frac{y^2}{b^2}+\frac{z^2}{c^2}=1$$

gives a locus which is the surface of an ellipsoid. Indeed in theory one could extend the number of dimensions to four or five or any larger number; and equations of the same form as those of the circle and sphere or of the ellipse and ellipsoid, but extended to include more coordinates, would define hyperspheres or hyperellipsoids.

Of course it is difficult to draw on a plane a diagram of a three-dimensional figure and impossible to draw one of a figure in four or more dimensions. However, the mathematics of these situations can still be developed. Such developments have application in current physics – for example, there are aspects of the theory of relativity that are studied with a model in which time is seen as something like a fourth dimension of space. One could perhaps imagine multidimensional biological models being explored for creatures whose movements were controlled to some extent by an acute sensitivity to some non-spatial modality, like sound or chemical concentration of a pheromone. But fortunately it is unnecessary to pursue the matter further here.

7

Equations and inequalities in a single variable

7.1 EQUATIONS, IDENTITIES AND RELATIONSHIPS

The mathematical symbol '=', well known to you though it is, can be very deceptive since it represents, according to circumstances, at least three distinct meanings. In a statement like

$$p=5 \tag{7.1}$$

the = means '(p) takes, of all possible values it might take, the value (5)'. This meaning can be elaborated as in statements like

$$2x^2+5x=\frac{x+90}{3} \tag{7.2}$$

where '=' now means or implies 'of all possible values the variable x might take, it is now actually taking a value which makes this statement true'. In fact equation (7.2) implies very directly that either $x=3$ or else $x=-6$, in both of which cases '=' is behaving in the same way as in equation (7.1). Statements like equations (7.1) and (7.2) are equations, which is what this chapter is about, so you expected nothing less.

But now consider the statement introduced in Chapter 4, Example 4(b))

$$\frac{1+\cos\theta}{\sin\theta}-\frac{\sin\theta}{1-\cos\theta}=0. \tag{7.3}$$

This statement is true for any real value of θ – any value whatever, positive or negative, rational or irrational. Here the '=' means just that: 'this statement is true whatever value the variable θ takes'. The statement in equation (7.3) is sometimes called an equation, but more properly it should be referred to as an identity (cf. section 2.5) and '=' should be replaced by '≡'. However, the use of '=' in this situation is well established and deeply rooted. It especially appears like this as an indication that an argument is proceeding which transforms an expression step by step without changing its value, thus

$$\begin{aligned} XYZ &= \ldots \\ &= \ldots \\ &= \ldots \qquad \text{QED.} \end{aligned}$$

In such a case, too, an identity is being asserted at each line – there is no equation.

Yet another example is provided by the functional relationship $y=f(x)$. In this case the '=' means '(y) takes up the position in its domain mapped by the function f which started from whatever position was taken up by x in *its* domain'. Alternatively, in this case when the function is being graphed, '=' means '(y) takes the ordinate value determined by the function f operating on the abscissa value x'. Again, '=' might revealingly be replaced by another symbol, perhaps '←' or '⇐', but here too the use of '=' is well established. In any case, a symbol like '⇐', indicating that $f(x)$ determines y in an explicit functional relationship, would be inappropriate for an implicit relationship such as that introduced at the beginning of chapter 6.

The examples I have given here are from the use of the symbol '=' in mathematical statements. The symbol is on no account to be confused with a similar-looking symbol '=' used by computer programmers. Programmers make statements like

$$x=x+1. \tag{7.4}$$

A computer can make sense of equation (7.4), but a human doing mathematics can't. If you move from a mathematics situation to a programming situation, be sure to change your symbols as well as your hat *en route*.

7.2 FIRST-DEGREE EQUATIONS WITH ONE UNKNOWN

Although other uses of the slippery '=' symbol will insinuate themselves, most of the remainder of this chapter is about equations. For completeness we can start with the simplest equations – those of the first degree with one unknown. With no great daring, we can call the unknown x. An equation which contains powers of x up to x^m (where m is a positive integer) is called an equation of the mth degree. So an equation of the first degree has $m=1$ and is of the form

$$Ax+B=0. \tag{7.5}$$

As you know, extracting a value of x from the equation is called solving it and the value extracted is called its root. Solving equation (7.5) certainly causes you no problem, but as a typical equation it provides an opportunity of listing what you can and can't do in solving equations.

It is commonly said that in dealing with an equation you can do anything to one side of the '=' sign provided that you do the same thing to the other. This is correct except that you are not allowed to divide each side by zero: dividing *anything* by zero is not allowed (cf. Section 2.4). (Here is an example of the sort of nonsense that can arise if you break this rule. Suppose that $b=a$. Then $ab=a^2$. Hence $ab-b^2=a^2-b^2$. Factorizing each side gives $b(a-b)=(a+b)(a-b)$ and dividing through by $a-b$ gives the result that $b=a+b$; but since $b=a$, it follows that $a=2a$ i.e. that $1=2$. The impossible result comes about as a consequence of dividing by $a-b$, which is zero.)

Care must also be taken in squaring each side. Because $(+p)^2=(-p)^2$, there is a risk, when squaring, of introducing a spurious solution to an equation. Although it wouldn't arise directly in solving an equation of the first degree, such an equation can be used to point to the trap. Thus if $x-a=0$, it is true that $x^2=a^2$, but it is not true that x could have the value of either $+a$ or $-a$.

7.3 EQUATIONS OF THE SECOND DEGREE (QUADRATIC EQUATIONS)

If $x-p=0$, then clearly $x=p$. If

$$(x-p)(x-q)=0 \tag{7.6}$$

then, by the property of zero, (equation (2.1)) one of the factors on the l.h.s. of equation (7.6) must be zero. That implies that either $x-p=0$ or $x-q=0$ so $x=p$ or $x=q$. It is perhaps worth stressing that at any time x has only one value, so the solution of equation (7.6) is not $x=p$ *and* $x=q$; an 'or' is part of the solution.

When the l.h.s. of equation (7.6) is multiplied out it leads to $x^2-(p+q)x+pq=0$. This is a slightly simplified version of the most general form of the l.h.s. which is

$$Ax^2+Bx+C=0 \tag{7.7}$$

where A, B, and C are positive or negative real numbers. You probably remember how to solve equation (7.7). The trick is to put it into the same form as equation (7.6) because equation (7.6) points directly to the solution. The method is called 'completing the square' and forces on to equation (7.7) the form or shape of the expansion of $(x+\alpha)^2$, i.e. the form of $x^2+2\alpha x+\alpha^2$. First divide through the equation by A (assuming that A is not zero) to give

$$x^2+\frac{B}{A}x+\frac{C}{A}=0.$$

Next complete the square involving the first two terms by writing the equation as

$$x^2+2\cdot\frac{B}{2A}x+\left(\frac{B}{2A}\right)^2-\left(\frac{B}{2A}\right)^2+\frac{C}{A}=0.$$

The first three terms of the equation now indeed have the form of $(x+\alpha)^2$ and this has been gained by adding zero (here disguised on the l.h.s. as $(B/2A)^2-(B/2A)^2$) to each side. Then genuinely add $(B/2A)^2-C/A$ to each side so that the equation becomes

$$\begin{aligned}\left(x+\frac{B}{2A}\right)^2&=\left(\frac{B}{2A}\right)^2-\frac{C}{A}\\&=\frac{B^2-4AC}{4A^2}.\end{aligned}$$

The square root of each side then gives either

$$x+\frac{B}{2A}=\frac{+\sqrt{B^2-4AC}}{2A}\qquad\text{or}\qquad x+\frac{B}{2A}=\frac{-\sqrt{B^2-4AC}}{2A}.$$

This means that the manoeuvres have enabled us to push equation (7.7) right back into the shape of equation (7.6), so we can write

$$\left(x+\frac{B}{2A}-\frac{\sqrt{B^2-4AC}}{2A}\right)\left(x+\frac{B}{2A}+\frac{\sqrt{B^2-4AC}}{2A}\right)=0,$$

from which the roots are available, as they were from equation (7.6). However, it is more usual to skip over that last step and say directly that the solution of equation (7.7) is

$$x=\frac{-B\pm\sqrt{B^2-4AC}}{2A}. \tag{7.8}$$

Here the symbol $\pm\sqrt{}$ is to be read as 'either $+\sqrt{}$ or $-\sqrt{}$'.

It is seen that an equation of the second degree has two roots. The quantity B^2-4AC is called the discriminant. Put P for $-B/2A$ and Q for $(B^2-4AC)/4A^2$. Then the two roots are $P+\sqrt{Q}$ and $P-\sqrt{Q}$. If the discriminant is positive (+ve), Q is +ve and the two roots are real. If the discriminant is zero, Q is also zero and the roots are real and equal (because they have the values P plus zero and P minus zero). If the discriminant is negative, we can write $Q=-R$ (where R is positive), in which case $\sqrt{-R}=\mathrm{i}\sqrt{R}$ and the two roots are $P+\mathrm{i}\sqrt{R}$ and $P-\mathrm{i}\sqrt{R}$; that is, the two roots are complex numbers which, furthermore, are complex conjugates. I shall return to these features of the roots of a quadratic equation in the next two sections.

Example 7(a) Solve

(a) $x^2-2x-15=0$;
(b) $45-13x-24x^2=0$;
(c) $x^2-2x+5=0$.

Answer: In each case apply the formula of equation (7.8).
(a) $A=1$, $B=-2$, $C=-15$, so the discriminant is $(-2)^2-4\cdot1\cdot(-15)=64$.

$$\therefore \qquad x=\frac{2\pm\sqrt{64}}{2}=\frac{2\pm8}{2}$$

Hence $x=5$ or $x=-3$.
(b) $A=-24$, $B=-13$, $C=45$, so the discriminant is $169+4\cdot24\cdot45=4489=67^2$.

$$\therefore \qquad x=\frac{13\pm67}{-48}$$

which leads to $x=-5/3$ or $x=9/8$.
(c) $A=1$, $B=-2$, $C=5$, and the discriminant is $4-20=-16=(\text{i}4)^2$.

$$\therefore \qquad x=\frac{2+\text{i}4}{2}$$

Hence $x=1+\text{i}2$ or $x=1-\text{i}2$.

EXERCISES 7A

1. Solve

   (a) $x^2-10x+24=0$;
   (b) $43x=35x^2+12$;
   (c) $x^2-4x+5=0$.

2. What value must be given to C to make the equation $16x^2-72x+C$ have equal roots?

7.4 GRAPHICAL SOLUTION OF EQUATIONS

If the first-degree *equation*, (7.5), is compared to the *function* $y=Ax+B$, two things can be noted. Firstly, the function when graphed as in Fig. 7.1 gives a straight line and for this reason equation (7.5) is called a linear equation in x. Secondly, the equation can be thought of as a cut or section of the graph representing the function, the cut being made by the line $y=0$ (which is the x-axis). Comparing the function with the standard equation for a straight line (equation (6.1)) we can recognize A as the gradient of the line

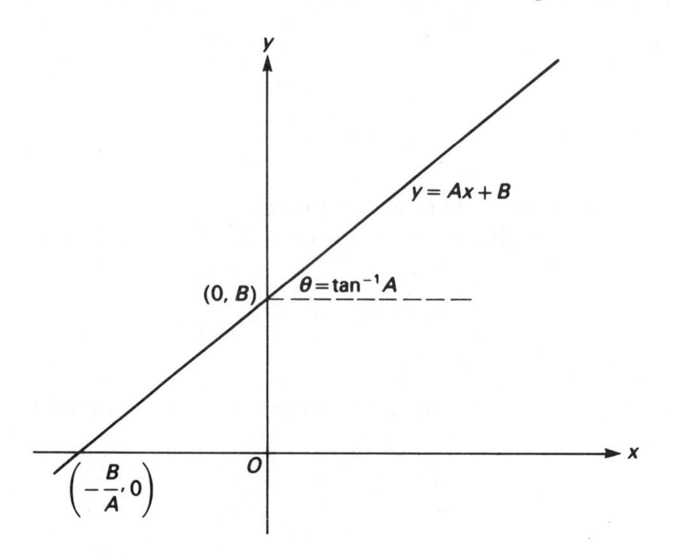

Figure 7.1

and B as the intercept on the y-axis. The intercept on the x-axis is $-B/A$ and this is the solution of the equation. That is, the solution of the equation is found at the intersection of the corresponding function with the x-axis. This account may seem to complicate the simple task of solving a first-degree equation, but its virtue is that it can be extended to the general statement: the solution to the equation $f(x)=0$ is given by the intersection of the x-axis, i.e. of $y=0$, with the graph of the function $y=f(x)$.

Applying the method to equations of the second degree, the graph to be examined is that given by the functional relationship $y=Ax^2+Bx+C$ which can be rearranged as

$$Ax^2+Bx-y+C=0. \tag{7.9}$$

Now compare this with the general second-degree functional relationship, equation (6.4)

$$ax^2+2hxy+by^2+2fx+2gy+c=0.$$

The comparison yields $a=A$, $f=B/2$, $g=-\frac{1}{2}$, $c=C$, and $h=b=0$. Although equation (7.9) provides the graph of a conic section, and $h=0$, it is not a circle, because $a \neq b$. In fact the condition realized is $h^2=ab$ which, by the rules stated in Chapter 6, shows that the graph is of a parabola. Indeed, it turns out to be a parabola with its axis parallel to the y-axis, at $x=-B/2A$ as shown in Fig. 7.2.

If the discriminant of the quadratic equation, (7.7), is positive, suppose that the real roots of the equation are α and β. In that case, the equation

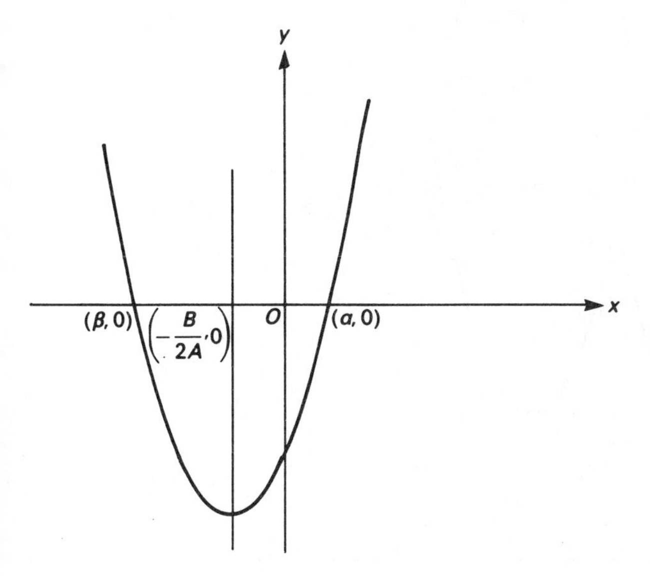

Figure 7.2

could be written, like equation (7.6) as $(x-\alpha)(x-\beta)=0$ and the corresponding function, of which the parabola in Fig. 7.2 would be the graph, is

$$y=(x-\alpha)(x-\beta). \tag{7.10}$$

Suppose that, of the two roots, α is the greater. Then the parabola cuts the x-axis (where $y=0$) at $x=\alpha$ and at $x=\beta$, with α lying to the right of β. When $x>\alpha$, both factors in equation (7.10) are positive and y is then positive. When $x<\beta$, both factors in equation (7.10) are negative so that y is again positive. But for $\beta<x<\alpha$, the first factor of equation (7.10) is negative, the second is positive and y is negative. All these conclusions appear immediately in the graph of Fig. 7.2 and hardly need analysis.

As the discriminant of equation (7.7) falls, i.e. as Q falls, where $x=P\pm\sqrt{Q}$, the points α and β move closer together. When the discriminant is zero, $\alpha=\beta$ and the two points at which the x-axis cuts the parabola coincide; the x-axis is then a tangent to the parabola, as in curve 1 of Fig. 7.3. When the discriminant becomes negative so that, if $Q=-R$, $x=P\pm i\sqrt{R}$, the x-axis no longer meets the parabola. Or rather, one should say that it no longer cuts it in real points, but it *does* cut it in two imaginary points. You can't see the points in Fig. 7.3, curve 2, because the points are imaginary and the plane of Fig. 7.3 is *not* the plane of the Argand diagram on which they might be plotted. One could, indeed, have a three-dimensional Argand diagram with the imaginary axis out of the plane of the paper and this might

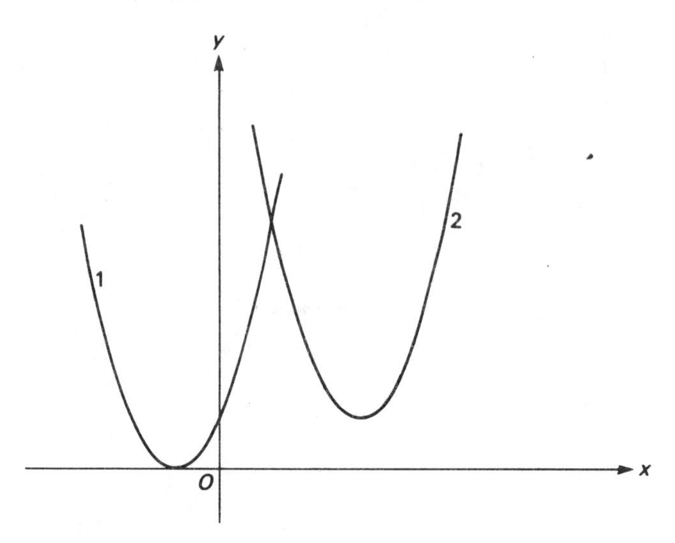

Figure 7.3

allow a sort of picture of the complex roots of the equation, but it makes too complicated a task for pursuing here.

If the equation to be solved is of the third degree, i.e. of the general form

$$Ax^3+Bx^2+Cx+D=0 \tag{7.11}$$

then the graph of the function $y=Ax^3+Bx^2+Cx+D$ has in principle the shape of the curves shown in Fig. 7.4. By 'in principle', I mean that the peak and trough may be flattened out to a greater or lesser extent, but usually can be discerned. In third-degree curves, if $x \to \infty$ then $y \to \pm\infty$ also, and equally, if $x \to -\infty$ then $y \to \mp\infty$ also. This is the case because, however much bigger B is than A, eventually a value of x is reached such that $|Ax^3|>|Bx^2|$. (Here there is a comparison of the magnitudes or moduli of Ax^3 and Bx^2; cf. Section 5.3). This has the result that the shape of the curve makes it certain that the line $y=0$, i.e. the x-axis, will cut the curve at least once. In fact, the x-axis may cut the curve three times (curve 1 of Fig. 7.4), in which case there are three real roots of equation (7.11), namely the values of x at L, M, and N in Fig. 7.4. Or the x-axis may cut the curve at one point and be a tangent to it at another (curve 2 of Fig. 7.4), in which case there are three real roots of equation (7.11), two of which are identical. Or, finally, the x-axis may meet the curve in only one point (curve 3 of Fig. 7.4) in which case the equation has one real root and two complex ones.

Going to equations of still higher degree, the number of varieties of graph becomes embarrassingly large. In the case of equations of the fourth degree, the graphs are parabola-like but possibly with a bump in the middle, allowing up to four real roots (curve 1 of Fig. 7.5); or there may be just two

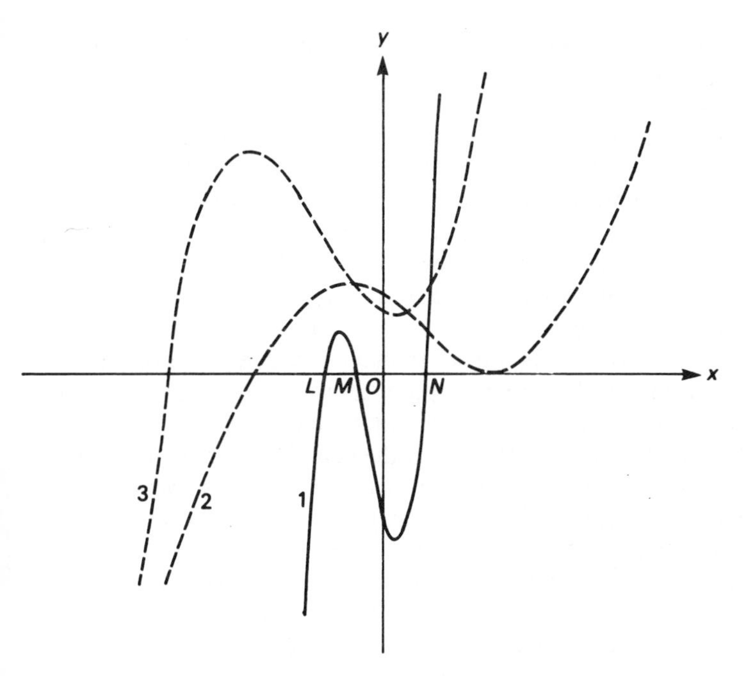

Figure 7.4

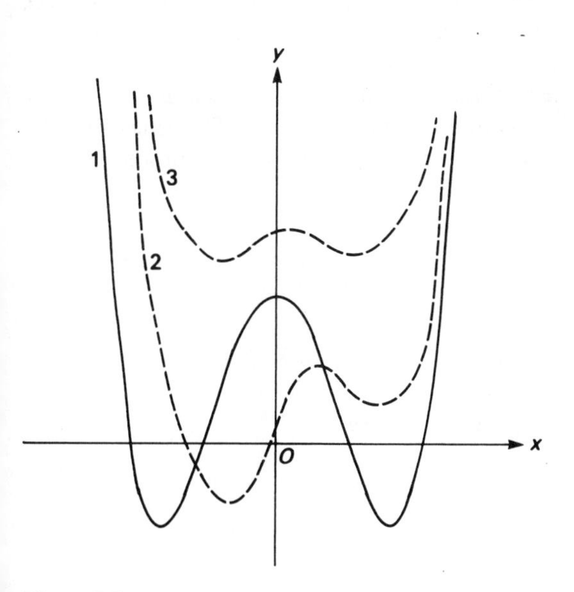

Figure 7.5

real roots or none at all (curves 2 and 3 of Fig. 7.5). In addition there are the various combinations with two or more coincident roots. Equations of the fifth degree involve graphs somewhat like the third degree (cubic) graphs of Fig. 7.4, showing that they may have one, three, or five, real roots. And for equations of higher odd-number degree the corresponding statements can be made: the number of real roots is 1, or 3, or 5, or .... The pattern is probably clear now: for equations of higher even number degree, the number of real roots is 0, or 2, or 4, or ....

Apart from allowing solutions of equations with powers of x, the graphical method also permits solution of equations which contain trigonometric functions, sometimes called transcendental equations. In many cases, a graphical method is the most practical way of obtaining a solution, and a graph may reveal an uncomfortably large number of answers. Consider, for example, the simple case of the equation $2\sin x=1$. A form of solution is $x=\sin^{-1}\frac{1}{2}$. But the solution to this equation is confined to the region $-\pi/2<x<\pi/2$ (cf. Section 4.5) and would be the value $\pi/6$ since $\sin\pi/6=\frac{1}{2}$ (cf. Section 4.6). However, since $\sin(\pi-\theta)=\sin\theta$ [from equation (4.6)], it follows that $\sin 5\pi/6=\frac{1}{2}$. In addition, $\sin(2\pi+\theta)=\sin\theta$ and then $\sin(2n\pi+\theta)=\sin\theta$ for any integer n, so that we end up with an infinite number of angles which satisfy the equation. This can be seen immediately in Fig. 7.6 where the curve of $\sin\theta$ reaches the value $\frac{1}{2}$, i.e. intersects the line $y=\frac{1}{2}$, an indefinitely large number of times both for positive and for negative values of x.

It nevertheless remains the case that any equation can be solved in respect of its real roots by graphical means. In particular cases there may be various kinds of short cut to the solution, but there has been space here only to illustrate the principle of the method. It is also to be noticed, however, that the graphical method yields only approximate solutions. This is because the drawing of a graph is essentially a practical experimental procedure and

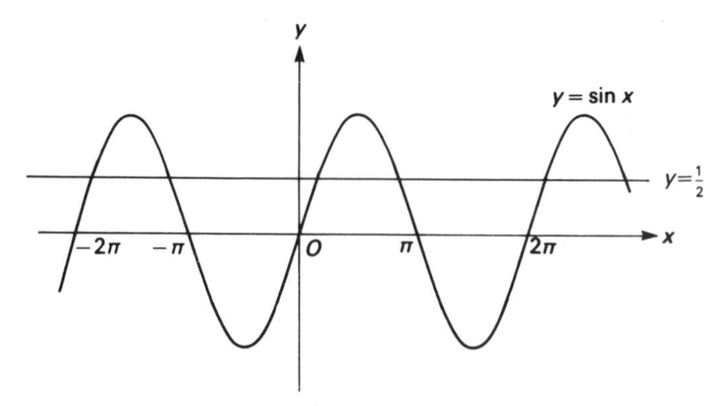

Figure 7.6

hence subject to error. Its accuracy is limited by, among other things, the thickness of the drawn lines.

Example 7(b) Solve graphically the equation $x^3+2x^2-4x-6=0$.

Answer: The function $y=x^3+2x^2-4x-6$ is graphed by the continuous line in Fig. 7.4 (though with the ordinate and abscissa scales suppressed). A larger-scale graph gives the values at which the graph cuts the x-axis (points L, M, and N in Fig. 7.4) as approximately -2.65, -1.20 and 1.85.

Example 7(c) Solve graphically the equation $3\cos x=2x-1$.

Answer: This equation can be solved by plotting the function

$$y=3\cos x-2x+1$$

and observing where it cuts the x-axis. However, it is simpler and more efficient to plot the two functions $y=3\cos x$ and $y=2x-1$. The former is the cosine function with an amplitude factor 3; the latter is a straight line, and hence very easy to graph. Where they meet, the equation is satisfied, and the value of x can be read off as in Fig. 7.7. The solution is approximately 1.134 radians (65°).

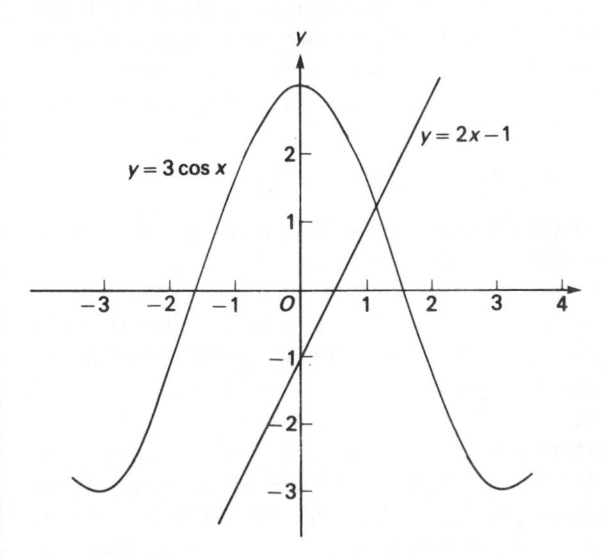

Figure 7.7

EXERCISES 7B

1. Find graphically the roots of the equation $x^4-5x^2+x+4=0$.

2. Show that the equation $\tan 2x = 2x + 2$ has an indefinitely large number of solutions and find the solution in the range $0 < x < \pi$.

7.5 POLYNOMIAL EQUATIONS

A polynomial equation is one of the form

$$a_n x^n + a_{n-1} x^{n-1} + \cdots + a_2 x^2 + a_1 x + a_0 = 0 \tag{7.12}$$

where the quantities a_i ($i = 0, 1, 2, \ldots, n$) are constant numbers called the coefficients of the corresponding powers of x. (Also, the l.h.s. of equation (7.12) is called a polynomial in x.) When $n = 2$, the polynomial is of degree 2 and equation (7.12) becomes a quadratic equation; and it becomes a cubic equation when $n = 3$ and the polynomial is of degree 3. The formula by which the solutions of a quadratic equation can be found is given by equation (7.8). A procedure has been worked out for solving a cubic equation, but it is complicated and too rarely needed to be worth repeating here. For equations of higher degree there are in general no formalized methods of solution; the roots must be found by graphical means or by numerical approximation.

Nevertheless, there are two important results about polynomial equations which are worth noting. The first is that a polynomial equation of degree n always has exactly n roots. This result has to some extent been implied in my account of the graphical solution of equations, but it is as well to state it explicitly. It means that, if $P(x)$ is a polynomial of degree n we could always write

$$P(x) = a_n(x - x_1)(x - x_2) \ldots (x - x_n)$$

where the polynomial has been divided into n factors like $(x - x_i)$ and x_1, $x_2, \ldots, x_n$ are the n roots of the equation $P(x) = 0$.

The second important result about polynomial equations is that, if all the coefficients a_i in equation (7.12) are real, then any roots of the equation which are complex numbers occur in pairs which are complex conjugates. The reason for this is easy to understand. If one root of $P(x) = 0$ is $\alpha + i\beta$ where α and β are real numbers then $(x - \alpha - i\beta)$ is a factor of $P(x)$, but if the product of all the factors is to end up with only real coefficients there must be at least one other factor which will remove all imaginary terms resulting from the imaginary part, $i\beta$, of this one. The only other factor which will do this job is the complex conjugate $(x - \alpha + i\beta)$, so another root of the equation is $\alpha - i\beta$.

These results reveal why the number of real roots of an equation of an odd-number degree is 1, or 3, or 5, or ..., and why the number of an equation of an even-number degree is 0, or 2, or 4, or .... And in doing so it also explains features of the graphs of polynomial functions which were noticeable in the previous section, such as the number of times the graphs cut the x-axis and the number of maxima and minima in the graphs.

7.6 INEQUALITIES

Inequalities are the relationships of 'greater than' ($>$) and 'less than' ($<$), as in statements like $x>3$ or $y<45$. Although complex numbers can be equal, one complex number cannot be greater than another. You can understand this from the Argand diagram representation. Among the real numbers, which lie on the x-axis of the Argand diagram, one can lie farther to the right of (be greater than) or farther to the left of (be less than) another. But such relativism has no meaning for points lying off the axes in the Argand plane. As a result, inequalities apply only to real numbers. But where the solution of an equation is generally a limited number of distinct roots, the solution of an inequality is a part of the domain of the variable.

Statements of inequality allow of solution, rather like equations, but with more restrictions of what you can do to the two sides. Numbers can be added to or subtracted from each side of an inequality. If a, b, and c are real numbers and $a>b$, then $a+c>b+c$ and $a-c>b-c$; but it is not necessarily true that $ac>bc$; this would be the case only if c is positive. If an inequality is multiplied through by a negative number, the inequality sign must be turned around. Thus when $a>b$, multiplying through by -1 makes $-a<-b$. Also, if $a>b$, $a^2>b^2$ only if $|a|>|b|$, to guard against b being negative. For example, $1>-2$ but $1^2 \ngtr (-2)^2$ (where $\ngtr$ means 'not greater than').

The various aspects of inequalities are best illustrated by examples.

Example 7(d) Solve $3x-1>\pi$.

Answer: $3x>\pi+1$, $\quad \therefore x>(\pi+1)/3$.

Example 7(e) Solve $x^2-7x+10>0$.

Answer: It follows that $(x-5)(x-2)>0$. Hence either (a) both factors are positive or (b) both factors are negative.

Case (a): $x-5>0$ and $x-2>0$ so that $x>5$ and $x>2$, i.e. $x>5$;
Case (b): $x-5<0$ and $x-2<0$ so that $x<5$ and $x<2$, i.e. $x<2$.

Hence the solution is $x>5$ or $x<2$.

Example 7(f) Solve $(x+1)/(x+2)>3$.

Answer: This is tricky. If the inequality is multiplied through by $x+2$, the multiplication may be by a negative quantity (if $x<-2$) and so will be against the rules. Instead write

$$\frac{x+1}{x+2}-3>0, \quad \text{i.e. } \frac{x+1-3x-6}{x+2}>0, \qquad \therefore \frac{2x+5}{x+2}<0.$$

But this means that the numerator and denominator of the fraction $(2x+5)/(x+2)$ are of opposite sign, i.e. either

(a) $2x+5>0$ and $x+2<0$, i.e. $x>-5/2$ and $x<-2$; or
(b) $2x+5<0$ and $x+2>0$, i.e. $x<-5/2$ and $x>-2$.

However, condition (b) is impossible, leaving $-5/2<x<-2$.

EXERCISES 7C

1. Solve $2x+5<x+7$.
2. Solve $-2<x^2+3x<10$.

8

Vectors, matrices and determinants

8.1 VECTORS AND SCALARS

In discussing complex numbers I wrote in Section 5.3 that in the Argand diagram a complex number is like a displacement. It is like a displacement, but it isn't actually a displacement. However, there is a mathematical quantity, namely a vector, which can very directly represent a displacement. A vector is a quantity which has magnitude and direction. A displacement, such as from A to B in Fig. 8.1 has magnitude (the length of the line AB) and direction ($A \rightarrow B$) and so can be represented by a vector. There are also many other quantities which have magnitude and direction and so are conveniently represented by vectors; examples are force, velocity and acceleration, electric and magnetic fields, and electric current. There are also

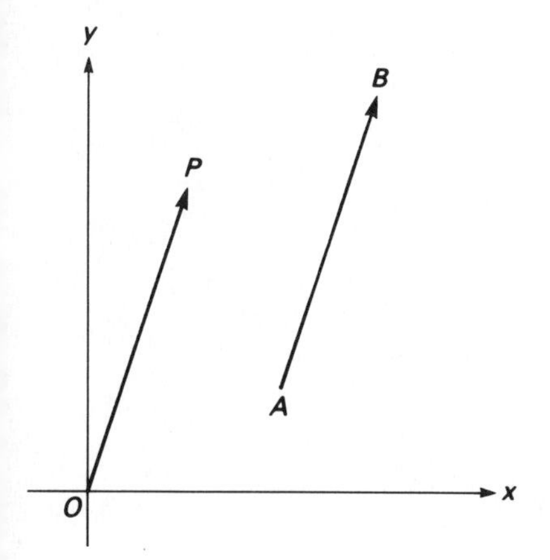

Figure 8.1

many quantities which have magnitude but no direction, such as numbers, time (which goes 'forward' only in a metaphorical sense), the mass of the Earth, the volume of beer drunk in England in 1991, the Financial Times index of Stock Exchange prices, etc. These quantities which have only magnitude are called scalars.

The displacement from A to B in Fig. 8.1 is the same in magnitude and direction as that from O to P. If the point of application is not specified, the vector is said to be free. Free vectors are referred to O, the origin of the Cartesian coordinates Ox and Oy in Figs 8.1 and 8.2. It has become conventional to symbolize vectors by characters in bold-face type and to symbolize the magnitude of the vector by the same character in ordinary type. So $\mathbf{V}$ might be a vector which is a displacement parallel to the direction OP in Fig. 8.2 and of magnitude V equal to the length OP. A vector in two dimensions, the plane of Fig. 8.2, can also be represented by a number pair. Thus the vector $\mathbf{V}$ in Fig. 8.2 could be specified as (p, q), the coordinates of P. In this case, we have for the magnitude that $V = \sqrt{p^2 + q^2}$. And the direction of the vector is given by θ, the angle between $\mathbf{V}$ and Ox, where $\theta = \tan^{-1} q/p$ and $-\pi < \theta < \pi$.

It is possible to add one vector to another and to add one scalar to another, but not to add a vector to a scalar. A scalar can multiply into another scalar (that's ordinary multiplication of numbers) and also into a vector (which would be the same as a vector multiplying into a scalar). But the multiplication of two vectors raises some problems: depending on the method of multiplication, the result is sometimes a vector and sometimes a scalar.

Example 8(a) What are the magnitude and direction of the vector (3, 7)?

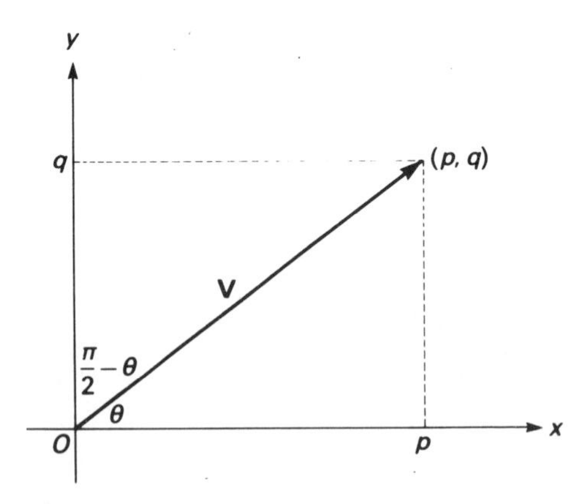

Figure 8.2

Answer: The magnitude is $\sqrt{3^2+7^2}=\sqrt{58}=7.62$. The direction is at an angle to the x-axis of $\tan^{-1} 7/3=66.8°$.

Example 8(b) Show that the vectors **P**, (4, 1), and **Q**, (−2, 8) are mutually perpendicular.

Answer: Let the direction of **P** and **Q** be θ_1 and θ_2 respectively. Then $\tan\theta_1=1/4$ and $\tan\theta_2=-8/2=-4$.

$\therefore \tan\theta_1 \cdot \tan\theta_2=-1$ (or, $\tan\theta_1=-\cot\theta_2$).

$\therefore \theta_1$ and θ_2 differ by $\pi/2$. (cf. Sections 4.7 and 6.1).

EXERCISES 8A

1. Find the vector which has magnitude 10 and is perpendicular to the vector (−9, 12).
2. Calculate the vector representing the displacement from the mid-point of A, (3, 7), and B, (−1, −3), to the mid-point of A, (3, 7), and C, (13, 45). What are its magnitude and direction? Show that it is parallel to and half the magnitude of the vector representing the displacement from B to C.

8.2 ADDITION OF VECTORS

It is usually said that vectors add by the parallelogram rule (sometimes called the parallelogram law), because the resultant of physically combining two displacements or two forces can most simply be reached by constructing a parallelogram like that in Fig. 8.3 which makes **R**, the diagonal of the parallelogram lying between the vectors **V** and **W**, the resultant of adding

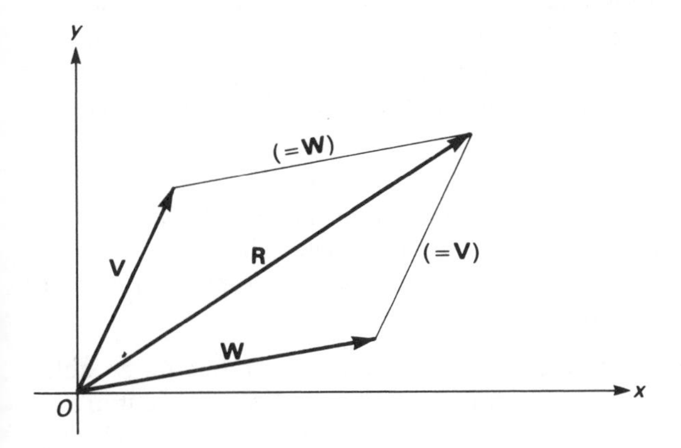

Figure 8.3

those two vectors. But writing the vectors as number pairs provides an easier approach to addition. The addition of two vectors, (l, m) and (r, s) is defined by:

$$(l, m)+(r, s)=(l+r, m+s).$$

This rule of addition is exactly the same as that for complex numbers and leads, as it did there, (cf. Section 5.4) to the parallelogram rule.

In this respect it is worth comparing complex numbers and vectors. If a complex number C is specified by (p, q), then p is the quantity of real number in C and q is the quantity of imaginary; that is, $C=p+iq$ where q must be multiplied by i and p is not multiplied by anything, but could be thought of, for the moment, as being multiplied by 1, the unity number of equation (2.1). If, however, the vector $\mathbf{V}$ were to be specified by the number pair (l, m) then l would be the quantity of x-directional units and m the quantity of y-directional units (where the units are of displacement, or force, or some other appropriate property). This can be made more explicit by writing $\hat{\mathbf{x}}$ for a vector of magnitude one unit in the x-direction and $\hat{\mathbf{y}}$ for a vector of magnitude one unit in the y-direction. With this notation the vector $\mathbf{V}=l\hat{\mathbf{x}}+m\hat{\mathbf{y}}$. If $\mathbf{W}$ is the vector (r, s) then the sum of $\mathbf{V}$ and $\mathbf{W}$ is $\mathbf{R}=(l+r)\hat{\mathbf{x}}+(m+s)\hat{\mathbf{y}}$, the meaning of which is clear (see Fig. 8.3).

The formalism of specifying planar vectors as number pairs extends without complication to dealing with vectors in three dimensions. In three dimensions vectors must, of course, be specified by number triplets, rather than by number pairs, with addition of the system of $\hat{\mathbf{z}}$, a vector of unit magnitude in the z-direction. Then if $\mathbf{J}$ is the vector (l, m, n) and $\mathbf{K}$ is the vector (r, s, t), the addition of the two is:

$$\mathbf{J}+\mathbf{K}=(l+r, m+s, n+t)=(l+r)\hat{\mathbf{x}}+(m+s)\hat{\mathbf{y}}+(n+t)\hat{\mathbf{z}}.$$

The vectors $\hat{\mathbf{x}}$, $\hat{\mathbf{y}}$, and $\hat{\mathbf{z}}$ are referred to as the unit vectors of the system. (In some books they are given as $\hat{\mathbf{i}}$, $\hat{\mathbf{j}}$, and $\hat{\mathbf{k}}$.) The magnitude, or modulus, of a three-dimensional vector (l, m, n) is $\sqrt{l^2+m^2+n^2}$. The magnitude of a vector $\mathbf{V}$ is sometimes written $|\mathbf{V}|$, i.e. the modulus of $\mathbf{V}$, as with complex numbers (cf. Section 5.3).

Example 8(c) If an object initially placed at the point P, $(1, 2, 3)$, is successively displaced by the three vectors specified by $\mathbf{A}$, $(4, 9, -3)$, $\mathbf{B}$, $(-5, 2, 2)$ and $\mathbf{C}$, $(3, -6, 4)$, what vector needs to be applied to shift the object back to its starting point?

Answer: In terms of displacements, a set of vectors operating on an object initially placed at the origin O, $(0, 0, 0)$, with the effect of moving it around and finally back to O would also move the object at P along parallel tracks but return it to P. The vectors $\mathbf{A}$, $\mathbf{B}$, and $\mathbf{C}$ are specified as free vectors, as if they were acting at O:

$$\mathbf{A}+\mathbf{B}+\mathbf{C}=(4-5+3,\ 9+2-6,\ -3+2+4)=(2,5,3)$$

and this is the position to which an object initially at O would have been moved by the successive application of **A** and then **B** and then **C**. The required vector, **D**, is such that $\mathbf{A}+\mathbf{B}+\mathbf{C}+\mathbf{D}$ is the vector $(0,0,0)$. Thus **D** is the vector $(-2,-5,-3)$.

Example 8(d) If **T** is the vector $(7,2)$, calculate the vector **S** which, when added to **T**, will give a resultant vector **R** in the direction of the x-axis and of magnitude 10.

Answer: Suppose the required vector **S** is (p,q). Then $\mathbf{R}=\mathbf{S}+\mathbf{T}=(p+7,q+2)$. If **R** is in the direction Ox, its y-component is zero and $q+2=0$, so that $q=-2$. If the magnitude of **R** is 10, $\sqrt{(p+7)^2+(q+2)^2}=10$; but $q=-2$ so $p=3$ and **S** is the vector $(3,-2)$.

EXERCISES 8B

1. A vector **A** given by $(1,2,3)$ can act on an object. Alternatively, the two vectors **B**, given by $(-3,-2,-1)$, and **C**, given by $(1,-1,1)$ can act on the object. Find a fourth vector, **D**, such that **D**, acting together with **B** and **C**, would have the same total effect as **A** acting alone.
2. Vectors **P** and **Q** are given by $(1,7)$ and $(5,5)$, respectively. Find the vector **R** such that $\mathbf{R}+\mathbf{P}$ has a magnitude of 64 and is perpendicular to **Q**.

8.3 MULTIPLICATION OF VECTORS

Product of a scalar and a vector

If **J** is the vector (l,m,n) and a is a scalar, then the product of a and **J** is $a\mathbf{J}=(al,am,an)$. In these circumstances the scalar is merely a magnifying factor and the product of the multiplication is a vector acting in the same direction as the original vector, **J**.

Scalar product of two vectors

If **J** and **K** are, respectively, the vectors (l,m,n) and (r,s,t) then they can form two products, a scalar product or a vector product. Suppose the vectors are inclined to each other at an angle α. Then the scalar product (sometimes colloquially called the dot product) is written $\mathbf{J}\cdot\mathbf{K}$ and

$$\mathbf{J}\cdot\mathbf{K}=lr+ms+nt=|\mathbf{J}|\cdot|\mathbf{K}|\cdot\cos\alpha. \qquad (8.1)$$

Either of the formulae of equation (8.1) can be proved from the other, but we need only the result, not the proof. Note that the name of this product comes about because the result is a scalar, not a vector. The scalar product is a maximum if the vectors lie in the same direction ($\alpha=0$) and is zero if the vectors lie perpendicular to each other ($\alpha=\pi/2$). The scalar product of a vector with itself is the square of its modulus, since for $\mathbf{J}\cdot\mathbf{J}$, as an example, $(l\hat{\mathbf{x}}+m\hat{\mathbf{y}}+n\hat{\mathbf{z}})\cdot(l\hat{\mathbf{x}}+m\hat{\mathbf{y}}+n\hat{\mathbf{z}})=l^2+m^2+n^2$.

Vector product of two vectors

This product is a vector lying in a direction perpendicular to both vectors being multiplied. It is written $\mathbf{J}\wedge\mathbf{K}$ (or $\mathbf{J}\times\mathbf{K}$) and is sometimes colloquially called the cross-product of the two vectors. Formulae for the vector product are:

$$\begin{gathered}\mathbf{J}\wedge\mathbf{K}=(mt-ns)\hat{\mathbf{x}}+(nr-lt)\hat{\mathbf{y}}+(ls-mr)\hat{\mathbf{z}}\\ |\mathbf{J}\wedge\mathbf{K}|=|\mathbf{J}|\cdot|\mathbf{K}|\cdot\sin\alpha.\end{gathered}\tag{8.2}$$

The vector product is maximum when the vectors being multiplied are perpendicular to each other ($\alpha=\pi/2$) and is zero if they are parallel ($\alpha=0$ or π); also $\mathbf{J}\wedge\mathbf{K}=-\mathbf{K}\wedge\mathbf{J}$.

I have stated the formulae for the various products as bald statements. To give even a semi-plausible justification for them goes further into physics or vector algebra than is needed for present purposes; so it is simplest to let equations (8.1) and (8.2) lie in place, mainly for reference. Note that the angle between two vectors can be calculated quite easily by use of the formulae in equation (8.1) for the scalar product. From the formulae it follows that

$$\alpha=\cos^{-1}\frac{\mathbf{J}\cdot\mathbf{K}}{|\mathbf{J}|\cdot|\mathbf{K}|}=\cos^{-1}\frac{lr+ms+nt}{\sqrt{(l^2+m^2+n^2)(r^2+s^2+t^2)}}$$

where $-\pi<\alpha<\pi$.

Example 8(e) What is the angle between the vectors $(\sqrt{3}, 1)$ and $(1, \sqrt{3})$?

Answer: There are two approaches to this question:

(a) the first vector is inclined to the x-axis at an angle

$$\theta_1=\tan^{-1}\frac{1}{\sqrt{3}}=\frac{\pi}{6}$$

and the second at

$$\theta_2=\tan^{-1}\frac{\sqrt{3}}{1}=\frac{\pi}{3}$$

(cf. Table 4.2). Thus $\theta_2-\theta_1=\pi/6$;

(b) $\theta_2-\theta_1=\cos^{-1}\dfrac{\sqrt{3}\cdot 1+1\cdot\sqrt{3}}{\sqrt{(3+1)(1+3)}}=\cos^{-1}\dfrac{2\sqrt{3}}{4}=\cos^{-1}\dfrac{\sqrt{3}}{2}=\dfrac{\pi}{6}.$

Example 8(f) Calculate the scalar and vector products of the two vectors in example 8(e).

Answer: The scalar product is $\sqrt{3}\cdot 1+1\cdot\sqrt{3}=2\sqrt{3}$; or alternatively is $\sqrt{4}\cdot\sqrt{4}\cdot\cos\pi/6=4\cdot\sqrt{3}/2=2\sqrt{3}$.

For the vector product, write the vectors as

$$\mathbf{A}=\sqrt{3}\hat{\mathbf{x}}+1\hat{\mathbf{y}}+0\hat{\mathbf{z}} \qquad \text{and} \qquad \mathbf{B}=1\hat{\mathbf{x}}+\sqrt{3}\hat{\mathbf{y}}+0\hat{\mathbf{z}}.$$

Then because **A** and **B** both lie in the x, y plane, their vector product must lie along the z-axis and have magnitude given by the second formula of equation (8.2) as $\sqrt{4}\cdot\sqrt{4}\cdot\sin\pi/6=2$. Alternatively, applying the first formula of equation (8.2)

$$\mathbf{A}\wedge\mathbf{B}=0\hat{\mathbf{x}}+0\hat{\mathbf{y}}+(\sqrt{3}\cdot\sqrt{3}-1\cdot 1)\hat{\mathbf{z}}=2\hat{\mathbf{z}}.$$

EXERCISES 8C

1. What is the angle between the vectors $(2, 2, -1)$ and $(-6, 0, 3)$, what are their magnitudes, and what are their scalar and vector products?
2. Show that the vectors $(3, 8, -2)$ and $(2, -1, -1)$ are perpendicular.

8.4 RESOLUTION OF A VECTOR

If a vector $\mathbf{V}=p\hat{\mathbf{x}}+q\hat{\mathbf{y}}$ has magnitude V and is inclined at an angle θ to the x-axis, as in Fig. 8.2, then $p=V\cos\theta$ and $q=V\sin\theta$. The quantities $p\hat{\mathbf{x}}$ and $q\hat{\mathbf{y}}$ are themselves vectors – $p\hat{\mathbf{x}}$ is the vector $(p, 0)$ – and are said to be the components of the vector **V** if the vector is resolved along the directions Ox and Oy. However, sometimes it is convenient to resolve a vector into two directions other than those of the Cartesian axes. Provided the two directions are at right angles, the results can still be said to be components of the vector, because the resultant of adding the two will be the original vector lying as the diagonal of the rectangle of which the two components form sides as shown in Fig. 8.4. Always, if a vector **V** is resolved in a direction at an angle ψ to itself, the component of the vector in that direction is $V\cos\psi$ and the component in the perpendicular direction is $V\sin\psi$. (This is also true of the components in the direction of the Cartesian axes, for if the vector makes an angle θ with the x-axis, then it makes an angle $\pi/2-\theta$ with the y-axis, so the component along the y-axis is $V\cos(\pi/2-\theta)=V\sin\theta$.)

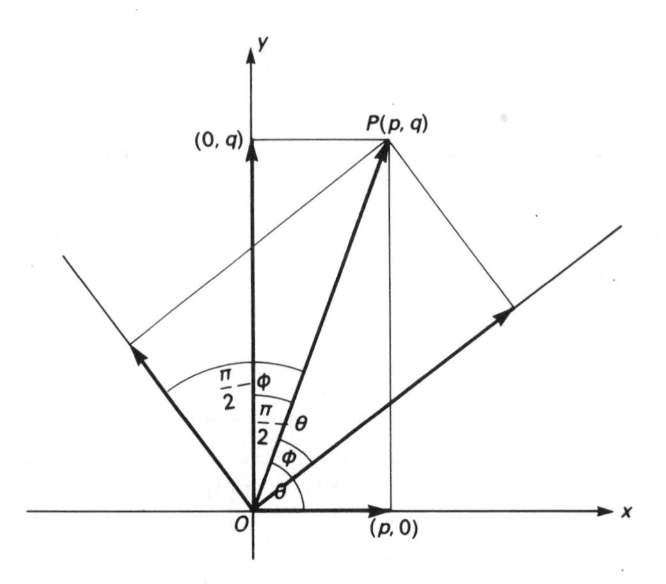

Figure 8.4

8.5 MATRICES

Consider the following situation. A membrane separates solutions of sodium chloride, NaCl, and potassium chloride, KCl, which have different concentrations on the two sides. The membrane is permeable to the ions which have concentrations $[Na^+]_i$, $[K^+]_i$, and $[Cl^-]_i$ where i is 1 or 2 for one side or the other of the membrane. Hydrostatic pressure is applied to side 1 of the membrane, causing water and ions to flow across to side 2, but the cations interfere with each other's transmembrane flow. The result is that the eventual concentration of sodium on side 2 of the membrane is proportional to 0.9 of the sodium on side 1, but reduced in proportion to 0.1 of the potassium concentration on side 1; and the eventual concentration of potassium on side 2 is proportional to 0.8 of the potassium on side 1, but reduced in proportion to 0.2 of the sodium on side 1. These complicated relationships can more simply be expressed as formulae:

$$\begin{aligned} 0.9[Na^+]_1 - 0.1[K^+]_1 &= [Na^+]_2 \\ -0.2[Na^+]_1 + 0.8[K^+]_1 &= [K^+]_2. \end{aligned} \tag{8.3}$$

I mean this model to imply that the calculations of equation (8.3) would give correct values for the concentrations on side 2 of the membrane whatever the starting concentrations on side 1. In that case, equation (8.3) provides a pattern of numbers which can be written as

$$\begin{pmatrix} 0.9 & -0.1 \\ -0.2 & 0.8 \end{pmatrix}$$

which act on or transform another pattern,

$$\begin{pmatrix} [Na^+]_1 \\ [K^+]_1 \end{pmatrix}$$

to produce a third pattern

$$\begin{pmatrix} [Na^+]_2 \\ [K^+]_2 \end{pmatrix}$$

The quantities in the big curved brackets are examples of the mathematical entities called matrices. Furthermore, equation (8.3) is an example of the kind of result obtained by applying a matrix in a practical situation; matrices can be of help in economically describing complicated phenomena in biology and physical chemistry.

I don't want to take matrix algebra very far – just far enough to show what matrices are and how they can appear in systems of equations. I shall therefore state rather tersely the rules for manipulating matrices, and refer back to these when matrices reappear at a later point. But note that many scientific matters (for example, in physical chemistry, topics such as symmetry of structures, and details of spectroscopy) are invariably discussed these days in terms of matrix operations.

8.6 MATRIX ALGEBRA

Definitions

A matrix is a rectangular array of numbers, called its elements. Thus a matrix with m rows and n columns would be written out as

$$\begin{pmatrix} a_{11} & a_{12} & a_{13} & \cdots & a_{1n} \\ a_{21} & a_{22} & a_{23} & \cdots & a_{2n} \\ a_{31} & & & & a_{3n} \\ \vdots & & & & \vdots \\ a_{m1} & a_{m2} & & \cdots & a_{mn} \end{pmatrix}$$

A matrix of m rows and n columns is said to be of order $m \times n$. Generally, the number of rows does not equal the number of columns. But if the two are equal, that is, if $m = n$, the matrix is said to be square.

If the matrix has $m=1$, it is reduced to $(a_{11}\ a_{12}\ a_{13} \ldots a_{1n})$ which is just like a vector except that n may be greater than 2 or 3 as in the two- and three-dimensional vectors we met earlier. In such a case the matrix is called a row vector. In a similar way, if $n=1$ the matrix is reduced to

$$\begin{pmatrix} a_{11} \\ a_{21} \\ \vdots \\ a_{m1} \end{pmatrix}$$

which again is like a vector written in an odd way; such a matrix is called a column vector.

Symbolism of matrix algebra

Matrices are commonly symbolized, like vectors, by letters printed in bold-face type. If necessary or appropriate, the numbers of rows and columns are indicated in subscripts. Thus, $\mathbf{M}$ might symbolize a matrix; and $\mathbf{M}_{p,q}$ (note the comma separating the letters p and q) would symbolize a matrix specifically of p rows and q columns. The elements of a matrix are normally written with the usual symbols of algebra. If the letters symbolizing the elements are given subscripts there is usually no comma but it is nevertheless to be understood that c_{ij} is the element in the ith row and jth column and it is sometimes referred to as the (i, j)th element. The arithmetic operations of addition, multiplication, etc. are indicated in the usual way. When it is necessary to write the matrix out in full, the elements are set out as an array enclosed between large curved brackets, as above, although some authors use large square brackets.

Matrix addition

If $\mathbf{A}_{m,n}$ is a matrix with m rows and n columns and $\mathbf{B}_{p,q}$ is a matrix with p rows and q columns, the two matrices can be added *only* if they are of exactly the same order, or in other words, of the same shape, so that $p=m$ and $q=n$. Suppose that a typical element of $\mathbf{A}_{m,n}$ can be written as a_{ij}, the element in the ith row and jth column. Similarly, the typical element of $\mathbf{B}_{p,q}$ can be written as b_{ij}. If it is indeed the case that $p=m$ and $q=n$ so that we can write $\mathbf{B}_{m,n}$ in place of $\mathbf{B}_{p,q}$, the matrices can be added to give, say, $\mathbf{C}_{m,n}=\mathbf{A}_{m,n}+\mathbf{B}_{m,n}$. Then the rule of matrix addition gives the typical element of $\mathbf{C}_{m,n}$ as

$$c_{ij}=a_{ij}+b_{ij}. \tag{8.4}$$

Matrix multiplication

As with vectors, a matrix can be multiplied by a scalar. If the scalar s multiplies the matrix $\mathbf{A}_{m,n}$ then the typical element of $s\mathbf{A}_{m,n}$ is sa_{ij}.

Two matrices, say $\mathbf{A}_{m,n}$ and $\mathbf{B}_{p,q}$, can be multiplied together *only* if the number of columns in the first matrix of the product equals the number of rows in the second matrix. Thus, if $\mathbf{C}=\mathbf{A}_{m,n}\times\mathbf{B}_{p,q}$, it is possible to find $\mathbf{C}$ only if $p=n$. The matrix $\mathbf{C}$ thus formed will have the same number of *rows* as the first matrix, $\mathbf{A}$, and the same number of *columns* as the second matrix, $\mathbf{B}$. So the only matrix multiplication which is permitted can be written as

$$\mathbf{C}_{m,q}=\mathbf{A}_{m,n}\times\mathbf{B}_{n,q}. \tag{8.5}$$

If the typical element of $\mathbf{C}_{m,q}$ is called c_{ij}, it is formed by taking the ith row of matrix $\mathbf{A}$ together with the jth column of matrix $\mathbf{B}$, multiplying the elements together in corresponding pairs and adding all the products. So

$$c_{ij}=a_{i1}b_{1j}+a_{i2}b_{2j}+a_{i3}b_{3j}+\cdots+a_{in}b_{nj}=\sum_{k=1}^{n} a_{ik}b_{kj}. \tag{8.6}$$

The process of matrix multiplication is as if the first matrix is turned clockwise on to its side and then pushed, column by column, into the second matrix.

If, for example, $\mathbf{A}_{2,2}=\begin{pmatrix}0.9 & -0.1\\ -0.2 & 0.8\end{pmatrix}$ and $\mathbf{B}_{2,1}=\begin{pmatrix}x\\ y\end{pmatrix}$ then the product $\mathbf{A}_{2,2}\times\mathbf{B}_{2,1}$ will be a matrix $\mathbf{C}$ with two rows (since $\mathbf{A}_{2,2}$ has two rows) and one column (since $\mathbf{B}_{2,1}$ has only one column). Suppose $\mathbf{C}_{2,1}=\begin{pmatrix}X\\ Y\end{pmatrix}$. Then

$$\begin{aligned}c_{11}=X&=a_{11}b_{11}+a_{12}b_{21}\\ &=0.9x-0.1y.\end{aligned} \tag{8.7a}$$

(To reach this line, imagine taking the first row of $\mathbf{A}$, $(0.9, -0.1)$, turning it clockwise so that it stands up as the column vector $\begin{pmatrix}0.9\\ -0.1\end{pmatrix}$, then multiplying it into B, $\begin{pmatrix}x\\ y\end{pmatrix}$, in paired elements, and finally turning the summed product anticlockwise before writing it down.) Similarly,

$$\begin{aligned}c_{21}=Y&=a_{21}b_{11}+a_{22}b_{21}\\ &=-0.2x+0.8y.\end{aligned} \tag{8.7b}$$

(To reach this line, imagine taking the second row of $\mathbf{A}$, $(-0.2, 0.8)$, and then turning and multiplying it into $\mathbf{B}$, as previously with the first row of $\mathbf{A}$.)

The statements in equations (8.7a) and (8.7b) are essentially the same as those of equation (8.3). This is because the rules of equations (8.5) and (8.6) have been chosen to give the results required in practical situations, so if the matrix $\mathbf{B}$ here represents the concentration of ions on one side of a membrane, the matrix $\mathbf{A}$ embodies the transforming property of the membrane system. The result, when $\mathbf{A}$ is multiplied into (i.e. operates on) $\mathbf{B}$ is the matrix **C** which represents the concentrations on the other side of the membrane.

8.7 ZERO AND UNIT MATRICES

Just as among the ordinary numbers there are the special numbers 0 and 1 (cf. section 2.3), so also there are corresponding special matrices.

Zero matrices

A matrix of whatever order of which all the elements are zero is called a zero matrix. Thus

$$\begin{pmatrix} 0 & 0 & 0 \\ 0 & 0 & 0 \end{pmatrix}$$

is a 2×3 zero matrix and

$$\begin{pmatrix} 0 & 0 & 0 \\ 0 & 0 & 0 \\ 0 & 0 & 0 \end{pmatrix}$$

is a 3×3 zero matrix. Write $\mathbf{Z}_{m,n}$ for the $m \times n$ zero matrix.

Unit matrices

In a square matrix of order $n \times n$ the diagonal of elements from the top left element, a_{11}, to the bottom right, a_{nn}, is called the leading diagonal. If, in such a matrix, all the elements in the leading diagonal, a_{ii}, for $i=1$ to n, have the value 1 and all other elements of the matrix have the value 0, it is called a unit matrix. Thus

$$\begin{pmatrix} 1 & 0 & 0 \\ 0 & 1 & 0 \\ 0 & 0 & 1 \end{pmatrix}$$

is a 3×3 unit matrix. Write $\mathbf{I}_{n,n}$ for the $n \times n$ unit matrix (although it is common practice, when the order of the matrix is clear from the context, to write a unit matrix just as **I**).

It is very easy to show (and you should do so for yourself as an exercise) that:

$$\mathbf{Z}_{m,n} + \mathbf{A}_{m,n} = \mathbf{A}_{m,n}$$
$$\mathbf{Z}_{m,n}\mathbf{A}_{n,p} = \mathbf{Z}_{m,p}$$
$$\mathbf{A}_{q,m}\mathbf{Z}_{m,n} = \mathbf{Z}_{q,n}$$
$$\mathbf{I}_{m,m}\mathbf{A}_{m,n} = \mathbf{A}_{m,n}\mathbf{I}_{n,n} = \mathbf{A}_{m,n}.$$

If consideration is limited to square matrices, suffixes can be omitted and you can write $\mathbf{ZA}=\mathbf{AZ}=\mathbf{Z}$, $\mathbf{A}+\mathbf{Z}=\mathbf{A}$, and $\mathbf{IA}=\mathbf{AI}=\mathbf{A}$.

8.8 WHAT ARE MATRICES?

It can be seen from the previous sections that matrices have some resemblances to numbers in the ways in which they can be handled. The commutative, associative, and distributive laws of arithmetic (cf. section 2.1) hold for the addition of matrices, i.e.

$$\mathbf{A}+\mathbf{B}=\mathbf{B}+\mathbf{A},$$
$$\mathbf{A}+(\mathbf{B}+\mathbf{C})=(\mathbf{A}+\mathbf{B})+\mathbf{C},$$
$$\mathbf{A}(\mathbf{B}+\mathbf{C})=\mathbf{AB}+\mathbf{AC}.$$

Also, **I** and **Z** play very similar roles in matrix algebra to those of 1 and 0 in ordinary arithmetic. However, the laws of arithmetic do not hold for the multiplication of matrices. Specifically, matrix multiplication is not commutative: usually,

$$\mathbf{AB}\neq\mathbf{BA}.$$

Therefore, an answer to the question 'What are matrices?' is that they are mathematical constructions in a heirarchy which starts with ordinary numbers (which themselves were elaborated into integers, rationals, and irrationals), then goes to number pairs, thence to longer strings of numbers (as in three-dimensional vectors), and with matrices develops into two-dimensional arrays. The hierarchy progresses still further, into three-dimensional arrays called tensors and into other constructions, but we have no need to go beyond two-dimensional arrays.

However, I introduced matrices a few pages ago in terms of patterns of numbers which might be connected with calculating the change of ionic concentration across a membrane. Although there are other uses and applications of matrices, it is appropriate to think of a matrix as a transforming operation. A matrix is typically used to operate on (multiply) a column vector to give another column vector; this was the manner in which, for the calculations given in equations (8.7a) and (8.7b), matrix **A** operated on column vector **B** to produce column vector **C** which is the transformed version of the original vector.

Example 8(g) Add the matrices

$$\mathbf{A}=\begin{pmatrix} 5 & -3 & 1 \\ 2 & 4 & 6 \\ 1 & 0 & -1 \end{pmatrix} \quad \text{and} \quad \mathbf{B}=\begin{pmatrix} 3 & 0 & 0 \\ 2 & -2 & 2 \\ 8 & 7 & 11 \end{pmatrix}.$$

Answer: The matrices are of identical size and shape and so may be added.

Use the addition rule of equation (8.4) to give

$$\mathbf{C}_{3,3}=\mathbf{A}_{3,3}+\mathbf{B}_{3,3}=\begin{pmatrix}5+3 & -3+0 & 1+0\\ 2+2 & 4-2 & 6+2\\ 1+8 & 0+7 & -1+11\end{pmatrix}=\begin{pmatrix}8 & -3 & 1\\ 4 & 2 & 8\\ 9 & 7 & 10\end{pmatrix}$$

Note that the same result would have been reached if we had taken $\mathbf{B}+\mathbf{A}$ instead of $\mathbf{A}+\mathbf{B}$. So matrices commute on addition.

Example 8(h) Multiply the matrices

$$\mathbf{A}=\begin{pmatrix}2 & 3\\ 1 & 0\\ 1 & -1\\ 3 & 2\end{pmatrix} \quad \text{and} \quad \mathbf{B}=\begin{pmatrix}8 & 5 & 2\\ 1 & 9 & 2\end{pmatrix}.$$

Answer: Matrix $\mathbf{A}$ has two columns and matrix $\mathbf{B}$ has two rows so the multiplication $\mathbf{C}=\mathbf{A}\times\mathbf{B}$ is permitted. The product will have four rows (like $\mathbf{A}$) and three columns (like $\mathbf{B}$). Use the multiplication rule of equation (8.6) to give

$$\mathbf{C}=\begin{pmatrix}2\times 8+3\times 1 & 2\times 5+3\times 9 & 2\times 2+3\times 2\\ 1\times 8+0\times 1 & 1\times 5+0\times 9 & 1\times 2+0\times 2\\ 1\times 8-1\times 1 & 1\times 5-1\times 9 & 1\times 2-1\times 2\\ 3\times 8+2\times 1 & 3\times 5+2\times 9 & 3\times 2+2\times 2\end{pmatrix}=\begin{pmatrix}19 & 37 & 10\\ 8 & 5 & 2\\ 7 & -4 & 0\\ 26 & 33 & 10\end{pmatrix}$$

Example 8(i) If

$$\mathbf{A}=\begin{pmatrix}1 & 0 & -1\\ 4 & 1 & 2\end{pmatrix} \quad \text{and} \quad \mathbf{B}=\begin{pmatrix}2 & 3\\ -1 & 1\\ 0 & 4\end{pmatrix} \text{ show that } \mathbf{A}\times\mathbf{B}\neq\mathbf{B}\times\mathbf{A}.$$

Answer: The numbers of rows and columns allow both products to be formed. Use the multiplication rules of equation (8.6)

$$\mathbf{A}\times\mathbf{B}=\begin{pmatrix}1 & 0 & -1\\ 4 & 1 & 2\end{pmatrix}\begin{pmatrix}2 & 3\\ -1 & 1\\ 0 & 4\end{pmatrix}=\begin{pmatrix}2+0+0 & 3+0-4\\ 8-1+0 & 12+1+8\end{pmatrix}=\begin{pmatrix}2 & -1\\ 7 & 21\end{pmatrix}$$

$$\mathbf{B}\times\mathbf{A}=\begin{pmatrix}2 & 3\\ -1 & 1\\ 0 & 4\end{pmatrix}\begin{pmatrix}1 & 0 & -1\\ 4 & 1 & 2\end{pmatrix}=\begin{pmatrix}2+12 & 0+3 & -2+6\\ -1+4 & 0+1 & 1+2\\ 0+16 & 0+4 & 0+8\end{pmatrix}=\begin{pmatrix}14 & 3 & 4\\ 3 & 1 & 3\\ 16 & 4 & 8\end{pmatrix}$$

Thus $\mathbf{A}\times\mathbf{B}\neq\mathbf{B}\times\mathbf{A}$.

This result could have been obtained without actually calculating the products. If $\mathbf{A}_{2,3} \times \mathbf{B}_{3,2} = \mathbf{C}$ then $\mathbf{C}$ must be a matrix with two rows (like $\mathbf{A}$) and two columns (like $\mathbf{B}$), whereas if $\mathbf{B}_{3,2} \times \mathbf{A}_{2,3} = \mathbf{D}$, then $\mathbf{D}$ is a matrix with three rows and three columns.

Conclusion: matrices are unlike numbers and do not necessarily commute on multiplication.

Example 8(j) If $\mathbf{V} = \binom{a}{b}$ is the vector $a\hat{\mathbf{x}} + b\hat{\mathbf{y}}$ and $\mathbf{M} = \begin{pmatrix} 0 & -1 \\ 1 & 0 \end{pmatrix}$, show that $\mathbf{M} \times \mathbf{V}$ gives a vector $\mathbf{W}$ which is perpendicular to but of the same magnitude as $\mathbf{V}$.

Answer:

$$\mathbf{W} = \begin{pmatrix} 0 & -1 \\ 1 & 0 \end{pmatrix} \begin{pmatrix} a \\ b \end{pmatrix} = \begin{pmatrix} 0-b \\ a+0 \end{pmatrix} = \begin{pmatrix} -b \\ a \end{pmatrix} = -b\hat{\mathbf{x}} + a\hat{\mathbf{y}}.$$

Suppose that $\mathbf{V}$ and $\mathbf{W}$ are inclined to the x-axis at angles θ_1 and θ_2, respectively. Then the gradient of $\mathbf{V}$ is $\tan\theta_1 = b/a$ and the gradient of $\mathbf{W}$ is $\tan\theta_2 = -a/b$. Thus $\tan\theta_1 \cdot \tan\theta_2 = -1$ so that θ_1 and θ_2 differ by $\pi/2$ (cf. section 4.7).

The magnitude of $\mathbf{V}$ is $V = \sqrt{a^2 + b^2}$. The magnitude of $\mathbf{W}$ is $W = \sqrt{(-b)^2 + a^2} = \sqrt{a^2 + b^2} = V$. Hence the operation of multiplying the matrix $\mathbf{M}$ into a vector rotates the vector through a right angle but leaves its size unchanged.

EXERCISES 8D

1. The matrix $\mathbf{R} = \begin{pmatrix} 1 & -1 \\ 1 & 1 \end{pmatrix}$ is multiplied into the vector $\mathbf{A} = \binom{a}{b} = a\hat{\mathbf{x}} + b\hat{\mathbf{y}}$, transforming it into the vector $\mathbf{B}$. $\mathbf{R}$ is then multiplied into $\mathbf{B}$, transforming it into $\mathbf{C}$. Compare the magnitudes of $\mathbf{A}$, $\mathbf{B}$, and $\mathbf{C}$. What are the relative directions of $\mathbf{A}$ and $\mathbf{C}$ and what can you conclude about the direction of $\mathbf{B}$? Test your conclusions by writing $\mathbf{A}$ as $\binom{1}{1}$ and then calculating and examining $\mathbf{B}$ and $\mathbf{C}$.
2. If $\mathbf{P} = \begin{pmatrix} 1 & 0 \\ 2 & -1 \end{pmatrix}$ and $\mathbf{Q} = \begin{pmatrix} 2 & -2 \\ 1 & -2 \end{pmatrix}$, prove that $(\mathbf{P}+\mathbf{Q})^2 = \mathbf{P}^2 + \mathbf{Q}^2$.

8.9 DETERMINANTS

A matrix is an array of elements and it defines an operation (e.g. a transformation). However, there is no value attached to a matrix. A matrix is something like a function. A function f as such has no value, although there may be a value if f operates on a quantity as in $y = f(x)$; if x takes the value x_1, then y is constrained to take a value y_1. Similarly, as we have seen, a matrix multiplied into (i.e. operating on) a vector results in another vector which has magnitude and direction. But although a matrix as such has no value, in the case of a square matrix there is a closely related object, called a **determinant** which does have a value.

As an example, a fourth-order determinant is a 4×4 array of numbers written as

$$\mathbf{D} = \begin{vmatrix} a_{11} & a_{12} & a_{13} & a_{14} \\ a_{21} & a_{22} & a_{23} & a_{24} \\ a_{31} & a_{32} & a_{33} & a_{34} \\ a_{41} & a_{42} & a_{43} & a_{44} \end{vmatrix}.$$

The value of the determinant is expressed as

$$\mathbf{D} = \sum (-1)^r a_{1i} a_{2j} a_{3k} a_{4l}. \tag{8.8}$$

The sum is of terms with various values of i, j, k, and l such that all possible combinations are taken of the products, using in each case only one element from each row and column. In this example, if the third element, a_{13}, were chosen from the first row, then in equation (8.8) $i=3$ so none of j, k, or l can be 3. If the second element were taken from the second row so that $j=2$ in equation (8.8), then k and l could only have the values 1 and 4 and a contribution to $\mathbf{D}$ might be the term $(-1)^r a_{13} a_{22} a_{34} a_{41}$. The value of r is the total number of *inversions* in the order of the columns in the term taking all columns into account (the rows all being in the correct order 1, 2, 3, 4). Going through all of them, $3 \to 2$ (i.e. a_{22} succeeding a_{13}) is an inversion; $3 \to 4$ is not; $3 \to 1$ is; $2 \to 4$ is not; $2 \to 1$ is; $4 \to 1$ is. So in all there are four inversions, $r=4$, $(-1)^r = +1$ and this particular term is positive.

Let us build up determinants from the beginning. The simplest determinant is of order 2 and we can call it $\mathbf{D}_2$ where $\mathbf{D}_2 = \begin{vmatrix} a & b \\ c & d \end{vmatrix}$. Then $\mathbf{D}_2 = ad - bc$. The term ad in $\mathbf{D}_2$ is positive because it involves no inversions; the term bc is negative because it involves one inversion.

Now look at

$$\mathbf{D}_3 = \begin{vmatrix} a & b & c \\ d & e & f \\ g & h & i \end{vmatrix}.$$

Two elements in $\mathbf{D}_3$ are $+aei$ and $-afh$, which can be found easily from inspection of $\mathbf{D}_3$. Picking out now the terms which contain b, we can extract from the arrangement of writing $\mathbf{D}_3$ that these are $-bdi$ and $+bfg$. The two final terms are $+cdh$ and $-ceg$. These six terms can be more succinctly written in pairs as $+a\begin{vmatrix} e & f \\ h & i \end{vmatrix}$, $-b\begin{vmatrix} d & f \\ g & i \end{vmatrix}$, and $+c\begin{vmatrix} d & e \\ g & h \end{vmatrix}$. In writing them out like this, it can be seen that a multiplies the determinant which is left after suppressing the row and column containing a in $\mathbf{D}_3$; the little determinant $\begin{vmatrix} e & f \\ h & i \end{vmatrix}$ is called the **minor** of a in $\mathbf{D}_3$. Similarly, $\begin{vmatrix} d & f \\ g & i \end{vmatrix}$ is the minor of b and

$\begin{vmatrix} d & e \\ g & h \end{vmatrix}$ is the minor of c. In expanding $\mathbf{D}_3$ as the sum of terms with products of elements of the first row and their minors, a plus sign goes in front of odd-numbered elements (the first and third, a and c) and a minus sign in front of the even-numbered elements (only the second, b, in this instance). Alternatively, $\mathbf{D}_3$ could have been written out as the sum of terms with the elements from the first column and their minors. So

$$\begin{aligned} \mathbf{D}_3 &= a\begin{vmatrix} e & f \\ h & i \end{vmatrix} - b\begin{vmatrix} d & f \\ g & i \end{vmatrix} + c\begin{vmatrix} d & e \\ g & h \end{vmatrix} \\ &= a\begin{vmatrix} e & f \\ h & i \end{vmatrix} - d\begin{vmatrix} b & c \\ h & i \end{vmatrix} + g\begin{vmatrix} b & c \\ e & f \end{vmatrix}. \end{aligned} \tag{8.9}$$

Or the expansion could have been in terms of any other row or column, provided that the minors were correctly assigned as +ve or −ve.

In the next chapter we shall use matrices and determinants in dealing with sets of equations. Here we shall merely work out the values of some determinants.

Example 8(k) Evaluate (a) $\mathbf{E}=\begin{vmatrix} 1 & 2 \\ 3 & 4 \end{vmatrix}$ and (b) $\mathbf{F}=\begin{vmatrix} 9 & 6 \\ 6 & 9 \end{vmatrix}$.

Answer: (a) $\mathbf{E}=1\cdot 4-2\cdot 3=-2$; (b) $\mathbf{F}=9^2-6^2=45$.

Example 8(l) Evaluate

$$\mathbf{G}=\begin{vmatrix} 2 & 3 & 0 \\ -1 & 4 & 2 \\ 1 & -1 & 2 \end{vmatrix}.$$

Answer:

$$\begin{aligned} \mathbf{G} &= 2\begin{vmatrix} 4 & 2 \\ -1 & 2 \end{vmatrix} - 3\begin{vmatrix} -1 & 2 \\ 1 & 2 \end{vmatrix} + 0 \\ &= 2(8-(-2))-3(-2-2)=20+12=32. \end{aligned}$$

Example 8(m) Evaluate

$$\mathbf{H}=\begin{vmatrix} 1 & 2 & 3 \\ 0 & 1 & 2 \\ 1 & 1 & 1 \end{vmatrix}.$$

Answer: In evaluating determinants, capitalize on the zeros, so expand $\mathbf{H}$

by its column elements with their minors as equation (8.9).

$$\mathbf{H}=1\begin{vmatrix}1 & 2\\ 1 & 1\end{vmatrix}-0+1\begin{vmatrix}2 & 3\\ 1 & 2\end{vmatrix}$$

$$=1(1-2)+1(4-3)=1-1=0.$$

EXERCISE 8E

Evaluate

(a) $\begin{vmatrix}5 & 1\\ 3 & 2\end{vmatrix}$

(b) $\begin{vmatrix}1 & 4 & 1\\ 2 & 0 & -2\\ 1 & 3 & 1\end{vmatrix}$

(c) $\begin{vmatrix}2 & 7 & 3\\ 1 & 3 & 5\\ 4 & 6 & -6\end{vmatrix}$

8.10 SIMPLIFYING DETERMINANTS

The trick in evaluating determinants is to simplify them as far as possible before setting out to make the calculation term by term. The main goal in simplification is to introduce as many zeros as possible and thereby to reduce the order of the determinant(s) which must be evaluated. The mechanics of simplification are embodied in three rules which are easy to prove, though it is unnecessary to do so here.

1. The value of a determinant is unaltered if a multiple of one row is added to or subtracted from another row or if a multiple of one column is added to or subtracted from another column.
2. If any two rows are interchanged or if any two columns are interchanged, the value of the determinant changes sign.
3. If each term in one row (or in one column) is multiplied by a factor, the value of the determinant is multiplied by that factor. Correspondingly, if each term in the row or column were divided by a factor, the value of the determinant would be divided by that factor.

Because, in the expansion of a determinant, every term contains factors from each row and each column, if any row or any column had all its elements zero, the value of the determinant would be zero. Hence a consequence of rule 1. is that if any one row (or column) is identical with another row (or column), or is a multiple of it, the value of the determinant is zero. This follows because an appropriate subtraction would then result in a line of zeros. It is also worth remembering that a determinant can be expanded by taking the elements of any row or column with their minors; the sign to be attached to a term is arrived at by starting with the first element of the first row as positive and then changing alternately negative, positive, negative, positive, element by element moving in any horizontal or vertical direction through the determinant (cf. example 8(o) below).

[In the answers to the following examples, Ri stands for the ith row and Cj stands for the jth column.]

Example 8(n) Evaluate

$$\mathbf{H}=\begin{vmatrix}1 & 2 & 3\\ 0 & 1 & 2\\ 1 & 1 & 1\end{vmatrix}.$$

(This repeats example 8(i).)

Answer: Take $R3-R1$ so

$$\mathbf{H}=\begin{vmatrix}1 & 2 & 3\\ 0 & 1 & 2\\ 0 & -1 & -2\end{vmatrix}.$$

Now $R3+R2$ gives only zeros in $R3$. Therefore $\mathbf{H}=0$.

Example 8(o) Evaluate

$$\mathbf{I}=\begin{vmatrix}3 & -1 & -1 & 2\\ 8 & -3 & 2 & 1\\ 2 & 3 & 4 & 5\\ 1 & 2 & 2 & 1\end{vmatrix}.$$

Answer:

$$R2-R4:\quad \mathbf{I}=\begin{vmatrix}3 & -1 & -1 & 2\\ 7 & -5 & 0 & 0\\ 2 & 3 & 4 & 5\\ 1 & 2 & 2 & 1\end{vmatrix}$$

$$C2+C1:\quad \mathbf{I}=\begin{vmatrix} 3 & 2 & -1 & 2 \\ 7 & 2 & 0 & 0 \\ 2 & 5 & 4 & 5 \\ 1 & 3 & 2 & 1 \end{vmatrix}; \qquad \text{then } C2-C4:\quad \mathbf{I}=\begin{vmatrix} 3 & 0 & -1 & 2 \\ 7 & 2 & 0 & 0 \\ 2 & 0 & 4 & 5 \\ 1 & 2 & 2 & 1 \end{vmatrix}$$

$$R4-R2:\quad \mathbf{I}=\begin{vmatrix} 3 & 0 & -1 & 2 \\ 7 & 2 & 0 & 0 \\ 2 & 0 & 4 & 5 \\ -6 & 0 & 2 & 1 \end{vmatrix}=2\begin{vmatrix} 3 & -1 & 2 \\ 2 & 4 & 5 \\ -6 & 2 & 1 \end{vmatrix}$$

[NB: the term is +ve because the element 2 is arrived at after one step right and one down from a_{11}]
Now in the third-order determinant

$$R3+2\cdot R1:\quad \mathbf{I}=2\begin{vmatrix} 3 & -1 & 2 \\ 2 & 4 & 5 \\ 0 & 0 & 5 \end{vmatrix}=2\times 5\begin{vmatrix} 3 & -1 \\ 2 & 4 \end{vmatrix}.$$

Hence $\mathbf{I}=10(12+2)=140$.

Example 8(p) Show that

$$\mathbf{J}=\begin{vmatrix} 1 & 1 & 1 \\ a & b & c \\ a^2 & b^2 & c^2 \end{vmatrix}=(c-a)(c-b)(b-a).$$

Answer:

$$C3-C2 \text{ and } C2-C1:\quad \mathbf{J}=\begin{vmatrix} 1 & 0 & 0 \\ a & (b-a) & (c-b) \\ a^2 & (b^2-a^2) & (c^2-b^2) \end{vmatrix}.$$

Now take the factor $(b-a)$ from the second column and the factor $(c-b)$ from the third, giving

$$J=(b-a)(c-b)\begin{vmatrix} 1 & 0 & 0 \\ a & 1 & 1 \\ a^2 & (b+a) & (c+b) \end{vmatrix}$$
$$=(b-a)(c-b)((c+b)-(b+a))$$
$$=(c-a)(c-b)(b-a).$$

EXERCISES 8F

1. Evaluate $\begin{vmatrix} 7 & 6 & -12 \\ 14 & -8 & 3 \\ -7 & 2 & -9 \end{vmatrix}$.

2. Show that $\begin{vmatrix} 1 & 1 & 1 & x \\ 1 & 1 & x & 1 \\ 1 & x & 1 & 1 \\ x & 1 & 1 & 1 \end{vmatrix} = (x-1)^3(x+3)$.

9

Equations with several unknowns

9.1 EQUATIONS IN TWO UNKNOWNS

As you know, two equations like

$$ax+by=c \tag{9.1a}$$

$$dx+ey=f \tag{9.1b}$$

form what is called a pair of simultaneous equations. The implication of the word 'simultaneous' is that there exists a pair of values x and y which satisfy both equations at the same time. This is not surprising since each of the formulae in equation (9.1) can be graphed as a straight line and the straight lines, if not parallel, intersect in just one point, and that point provides a pair of coordinates, (x, y), which satisfy both equations.

If equation (9.1) is solved in a formal way, the method is to eliminate one variable, say y, leaving a single equation in x; to solve this for x; and to substitute the resulting value of x back into one of the equations so as to obtain a value for y. Thus equation (9.1b) gives $y=(f-dx)/e$. Putting this into equation (9.1a) gives

$$ax+\frac{bf-bdx}{e}=c.$$

Multiply through by e and rearrange to get

$$aex-bdx=ce-bf.$$

$$\therefore \qquad x=\frac{ce-bf}{ae-bd}.$$

Put this value back into equation (9.1a) to get

$$\frac{ace-abf}{ae-bd}+by=c.$$

Multiply through by $(ae-bd)$ to get

$$ace-abf+abey-b^2dy=ace-bcd.$$

This rearranges to

$$b(ae-bd)y=b(af-cd)$$

so that

$$y=\frac{af-cd}{ae-bd}.$$

This method yields a perfectly good answer – that is to say, an answer which is correct and quite efficiently obtained. But the method is limited in that it becomes steadily more complicated as the number of equations increases to accommodate three or more unknowns. So a more useful variant of it, which is very often used, is to multiply equation (9.1a) by e (the coefficient of y in equation (9.1b)) and to multiply equation (9.1b) by b (the coefficient of y in equation (9.1a)). This gives

$$aex+bey=ce \tag{9.2a}$$

$$bdx+bey=bf. \tag{9.2b}$$

Because the l.h.s. and the r.h.s. of equation (9.2b) are exactly equal, we can manipulate equation (9.2a) by taking the l.h.s. of equation (9.2b) from its left side and the r.h.s. of equation (9.2b) from its right side and still be left with an equality, namely,

$$(ae-bd)x=ce-bf.$$

The terms in y have been eliminated and, as before, $x=(ce-bf)/(ae-bd)$. A similar procedure gives the same answer as before for y, which is what you would expect.

Now examine the values reached for x and y. First note that the two fractions have the same denominator, $ae-bd$. Then see that this denominator can be written as a second-order determinant

$$\Delta=\begin{vmatrix} a & b \\ d & e \end{vmatrix}. \tag{9.3}$$

Now note that the pattern of elements in the determinant is the same as the pattern of coefficients of x and y on the l.h.s. of equation (9.1).

In the case of the fraction giving the value of x, the numerator, $ce-bf$, can be written as a determinant

$$\Delta_1=\begin{vmatrix} c & b \\ f & e \end{vmatrix}. \tag{9.4}$$

Similarly, the numerator of the fraction for y can be written as the determinant

$$\Delta_2 = \begin{vmatrix} a & c \\ d & f \end{vmatrix}. \tag{9.5}$$

Now look for the connection between the determinants in equations (9.3)–(9.5). That in equation (9.3), labelled Δ, is the pattern of the x, y coefficients in the original equations; that in equation (9.4), the numerator of the x-fraction, is nearly the same, but has the r.h.s. coefficients of equation (9.1) substituted for the 'x' or first column of equation (9.3) and for that reason it has been labelled Δ_1. Finally, the determinant in equation (9.5), the numerator of the y-fraction, is nearly the same again, but with the r.h.s. coefficients of equation (9.1) now substituted in the 'y' or second column, for which reason it has been labelled Δ_2. Thus the solutions of equation (9.1) can be succinctly written as

$$x = \frac{\Delta_1}{\Delta} \quad \text{and} \quad y = \frac{\Delta_2}{\Delta}. \tag{9.6}$$

Before carrying this method further, it is useful to examine an obstacle to its employment. If it should happen that $\Delta = 0$, the operation of dividing by Δ would not be allowed and the 'solutions' expressed in equation (9.6) would have no meaning. In what circumstances does this situation arise? By definition, $\Delta = \begin{vmatrix} a & b \\ d & e \end{vmatrix} = ae - bd$. So $\Delta = 0$ means that $ae = bd$ so that $e/d = b/a$. Look back to the original simultaneous equations (9.1a) and (9.1b) regarded as equations of straight lines. The straight line of equation (9.1a) has a gradient (if the equation is pushed into the form $y = mx + c$) of $-b/a$; and the straight line of equation (9.1b) has a gradient of $-e/d$. Now it is apparent what could prevent solution of the equations: $\Delta = 0$ if the two gradients are equal and the lines are parallel so that they do not, after all, intersect in a point with coordinates which give the solution. (A variant of this is that the lines are not just parallel but are identical, so that they 'meet' at all points along their lengths, but let us suppose that this circumstance would be noticed before an attempt was made to solve the equations.)

Finally, it is worth making the point that although the solution of equation (9.1) written with determinants as in equation (9.6) is a good route to the correct answer (provided $\Delta \neq 0$), there may be particular cases where substitution for x or y or elimination of one of them is a preferable way to take. Each problem should be judged on its merits.

Example 9(a) Solve for x and y:

$$2x - 3y = 13$$
$$4x + y = 5.$$

Answer: Using determinants,

$$\Delta=\begin{vmatrix}2 & -3\\4 & 1\end{vmatrix}=2+12=14$$

$$\Delta_1=\begin{vmatrix}13 & -3\\5 & 1\end{vmatrix}=13+15=28$$

$$\Delta_2=\begin{vmatrix}2 & 13\\4 & 5\end{vmatrix}=10-52=-42$$

Hence,

$$x=\frac{\Delta_1}{\Delta}=\frac{28}{14}=2; \qquad y=\frac{\Delta_2}{\Delta}=\frac{-42}{14}=-3.$$

Example 9(b) Solve for x and y:

$$2x-y=0$$
$$5x-2y=2.$$

Answer: Because of the simplicity of the first equation, use the method of substitution. But substitute for y; substituting for x would introduce awkward factors of $\frac{1}{2}$ and risk error. Thus substituting $y=2x$ in the second equation gives $5x-4x=2$. So $x=2$ and therefore $y=4$.

Example 9(c) Solve for x and y:

$$y=3x$$
$$9x=3y-2.$$

Answer: The lines are parallel, the equations are insoluble.

$$\Delta=\begin{vmatrix}3 & -1\\9 & -3\end{vmatrix}=0.$$

EXERCISE 9A

Solve for x and y:

(a) $4x-7y=30$ and $x+3y=17$;
(b) $x-y=1$ and $5x+7y=5$;
(c) $12y=8x+5$ and $2x-3y=11$.

9.2 OVER- AND UNDER-DETERMINATION OF UNKNOWNS

A pair of equations with two unknowns, x and y, as in equation (9.1), can be graphed as a pair of straight lines which meet in a point (assuming they are not parallel). Three equations with two unknowns could be graphed as three straight lines. These lines might all meet in a point, in which case the point would have been uniquely specified by any two of the lines and hence is over-determined by the third line, which contributes unneeded information to the problem. Alternatively, the lines might not have a common crossing point, but instead meet in pairs; and in this case the problem is over-determined to the level of inconsistency.

At the other extreme, if only one equation is given, graphing as just one straight line, it contains indeed the point where the second line would have cut it if the second line were there; but it also contains an infinity of other points where other lines might have cut it and there is no reason for giving any one of these points precedence over the others.

Therefore, if equations connect two unknowns, just two equations are needed to specify them uniquely and neither more nor less than two. This argument can be extended to sets of equations with more than two unknowns. If there are k unknowns, their specification requires just k equations; fewer than k under-determines them and more than k over-determines.

9.3 SIMULTANEOUS LINEAR EQUATIONS IN MANY UNKNOWNS (CRAMER'S RULE)

If there are n unknowns, called $x_1, x_2, \ldots, x_n$, a typical linear equation connecting them would be

$$a_1x_1 + b_1x_2 + c_1x_3 + \cdots + n_1x_n = y_1.$$

There are n terms on the l.h.s. (although one or more of the coefficients – a_1, b_1, etc. – may be zero); and to have the unknowns neither over- nor under-determined there will be just n equations like this one. The second of them might be

$$a_2x_1 + b_2x_2 + \cdots + n_2x_n = y_2,$$

the third

$$a_3x_1 + b_3x_2 + \cdots + n_3x_n = y_3,$$

and so on. If all the equations are written out like this, the coefficients will always form a square $n \times n$ array. In turn, this means that they can always be the elements of a determinant. Following the pattern found in looking at the solution of a pair of equations with just two unknowns, call this determinant Δ.

To solve the equations, the pattern found with the pair of equations is followed further. Put Δ_1 for the determinant formed when the column $\begin{pmatrix} y_1 \\ y_2 \\ \vdots \\ y_n \end{pmatrix}$ is substituted for the column $\begin{pmatrix} a_1 \\ a_2 \\ \vdots \\ a_n \end{pmatrix}$ in Δ. In the same way, the determinant Δ_2 is formed when the column of y elements is substituted for the column of b elements in Δ, and so on for the other determinants, Δ_i. The solution of the set of equations is then

$$x_1 = \frac{\Delta_1}{\Delta}; \qquad x_2 = \frac{\Delta_2}{\Delta}; \qquad \dots \qquad x_n = \frac{\Delta_n}{\Delta}.$$

The method by which the solution is obtained here is called Cramer's rule.

Example 9(d) Solve for x, y, and z:

$$\begin{aligned} 2x + \;\; y + 2z &= 5 \\ x - 5y + \;\; z &= 8 \\ 4x + 2y - \;\; z &= 0. \end{aligned}$$

Answer: Using Cramer's rule with the notations used before,

$$\Delta = \begin{vmatrix} 2 & 1 & 2 \\ 1 & -5 & 1 \\ 4 & 2 & -1 \end{vmatrix}; \qquad C1-C3\text{:} \quad \Delta = \begin{vmatrix} 0 & 1 & 2 \\ 0 & -5 & 1 \\ 5 & 2 & -1 \end{vmatrix} = 5\begin{vmatrix} 1 & 2 \\ -5 & 1 \end{vmatrix}$$

$$= 5(1+10)$$

$$\therefore \; \Delta = 55.$$

$$\Delta_1 = \begin{vmatrix} 5 & 1 & 2 \\ 8 & -5 & 1 \\ 0 & 2 & -1 \end{vmatrix};$$

$$C2+2C3\text{:} \quad \Delta_1 = \begin{vmatrix} 5 & 5 & 2 \\ 8 & -3 & 1 \\ 0 & 0 & -1 \end{vmatrix} = -1\begin{vmatrix} 5 & 5 \\ 8 & -3 \end{vmatrix} = -1(-15-40)$$

$$\therefore \; \Delta_1 = 55.$$

$$\Delta_2 = \begin{vmatrix} 2 & 5 & 2 \\ 1 & 8 & 1 \\ 4 & 0 & -1 \end{vmatrix}; \qquad C1-C3: \ \Delta_2 = \begin{vmatrix} 0 & 5 & 2 \\ 0 & 8 & 1 \\ 5 & 0 & -1 \end{vmatrix} = 5\begin{vmatrix} 5 & 2 \\ 8 & 1 \end{vmatrix} = 5(5-16)$$

$\therefore \ \Delta_2 = -55.$

$$\Delta_3 = \begin{vmatrix} 2 & 1 & 5 \\ 1 & -5 & 8 \\ 4 & 2 & 0 \end{vmatrix};$$

$$R3-2R1: \ \Delta_3 = \begin{vmatrix} 2 & 1 & 5 \\ 1 & -5 & 8 \\ 0 & 0 & -10 \end{vmatrix} = -10\begin{vmatrix} 2 & 1 \\ 1 & -5 \end{vmatrix} = -10(-10-1)$$

$\therefore \ \Delta_3 = 110.$

Hence,

$$x = \frac{\Delta_1}{\Delta} = 1; \qquad y = \frac{\Delta_2}{\Delta} = -1; \qquad z = \frac{\Delta_3}{\Delta} = 2.$$

Example 9(e) Solve for a, b, and c:

$$\begin{aligned} 3a - \ b + 3c &= 7 \\ 5a + 4b + 5c &= 3 \\ a + 2b + \ c &= 6. \end{aligned}$$

Answer:

$$\Delta = \begin{vmatrix} 3 & -1 & 3 \\ 5 & 4 & 5 \\ 1 & 2 & 1 \end{vmatrix};$$

$C1 = C3$, so $\Delta = 0$; hence the equations are not soluble.

EXERCISES 9B

1. Solve for x, y, and z:

$$\begin{aligned} 2x + 3y + \ 4z &= 1 \\ x - \ y + \ 8z &= 1 \\ 3x + \ y + 12z &= 1. \end{aligned}$$

2. Solve for w, x, y, and z:

$$\begin{aligned} 2w+x-y-2z&=5\\ 3x+4y+z&=2\\ 5w+3x+2y&=5\\ 3w+2y+2z&=1. \end{aligned}$$

9.4 CURVE FITTING WITH MINIMAL DATA

It is often necessary in scientific work to fit experimental points to a curved or straight line. The best circumstances in which to do this are those in which very many experimental points are available and theory or informed guesswork gives a clue to the shape of the relationship between the independent variable, x say, which is typically set in an experiment, and the dependent variable, y say, the quantity appearing as the measured experimental result. The problem is then one in statistics which we meet later in Chapter 14. However, it is the opposite extreme I want to look at here – the situation, often arising in practice, of having few experimental results with which to establish or check a hypothesis.

So suppose that experimentally a quantity x is set and another quantity y is measured. Look first at the simplest relationship there might be between x and y – a straight line. We know that the equation of a straight line can always be pushed into the form $y=mx+c$ (cf. Section 9.1). Appreciate that the unknowns now are m and c, not x and y, and since there are two unknowns, so two equations will be required to specify them uniquely. The two equations which will allow m and c to be calculated come from two experimental pairs, (x_1, y_1) and (x_2, y_2), which provide the simultaneous equations in m and c:

$$\begin{aligned} mx_1+c&=y_1\\ mx_2+c&=y_2. \end{aligned}$$

Another way of looking at this situation is to realize that if only one point, (x_1, y_1), is placed between the axes in the x, y plane, infinitely many straight lines could be drawn through it, but as soon as a second point, (x_2, y_2), is added, the two points uniquely define just one straight line – that which joins them and extends on either side. If the pairs of numbers, (x_1, y_1) and (x_2, y_2), indeed come as experimental results, a third experiment would yield another pair, (x_3, y_3), which would be most unlikely, because of experimental error, to lie exactly on the straight line established by the other two. This would be a typical case in which m and c were over-determined by excessive inconsistent data.

Two pairs of (x, y) values are necessary and sufficient to specify a linear equation (straight line). Three points similarly will specify a quadratic equation (parabola), four points a cubic, and so on. Thus in general the experimental results will be able to specify a polynomial equation like

$$y=a_0+a_1x+a_2x^2+a_3x^3+\cdots+a_nx^n.$$

The unknowns here are the quantities a_0, a_1, a_2, etc, so $n+1$ experimental (x, y) pairs will allow all of them to be determined.

Looking more closely at the case of three experimental points used to establish a parabolic relation between y and x, the three equations will be:

$$a_0+a_1x_1+a_2x_1^2=y_1$$
$$a_0+a_1x_2+a_2x_2^2=y_2$$
$$a_0+a_1x_3+a_2x_3^2=y_3.$$

In this set, *the coefficients of the unknowns* are the quantities 1, x_1, x_2^2, etc. Thus the determinant formed of the array of coefficients is

$$\Delta=\begin{vmatrix} 1 & x_1 & x_1^2 \\ 1 & x_2 & x_2^2 \\ 1 & x_3 & x_3^2 \end{vmatrix}$$

and comparing this with Example 8(p) it is straightforward to show that $\Delta=(x_2-x_1)(x_3-x_2)(x_3-x_1)$. Unless experimental results have produced particularly convenient combinations of x and y, the determinants Δ_i $(i=1, 2, 3)$ will not reduce to simple expressions like this one; but, by use of the procedures explained earlier the constants $a_0, a_1, \ldots$ can be obtained.

Example 9(*f*) Find the parabola of the form $y=a_0+a_1x+a_2x^2$ which contains the points (1, 5), (3, −3) and (5, −3). Which values of x would make $y=0$?

Answer: Substituting the x, y values successively gives three simultaneous equations, for which

$$\Delta=\begin{vmatrix} 1 & 1 & 1 \\ 1 & 3 & 9 \\ 1 & 5 & 25 \end{vmatrix}=(3-1)(5-3)(5-1), \quad \text{by the above formula}=16.$$

$$\Delta_1=\begin{vmatrix} 5 & 1 & 1 \\ -3 & 3 & 9 \\ -3 & 5 & 25 \end{vmatrix}; \quad \begin{matrix} C3-3C2 \\ \text{and} \\ C2+C1 \end{matrix} \quad \Delta_1=\begin{vmatrix} 5 & 6 & -2 \\ -3 & 0 & 0 \\ -3 & 2 & 10 \end{vmatrix}=+3\begin{vmatrix} 6 & -2 \\ 2 & 10 \end{vmatrix}$$

$$=3\times 64=192.$$

$$\Delta_2 = \begin{vmatrix} 1 & 5 & 1 \\ 1 & -3 & 9 \\ 1 & -3 & 25 \end{vmatrix}; \qquad \begin{matrix} R3-R2 \\ \text{and} \\ R2-R1 \end{matrix} \qquad \Delta_2 = \begin{vmatrix} 1 & 5 & 1 \\ 0 & -8 & 8 \\ 0 & 0 & 16 \end{vmatrix} = +16\begin{vmatrix} 1 & 5 \\ 0 & -8 \end{vmatrix}$$

$$= -8 \times 16 = -128.$$

$$\Delta_3 = \begin{vmatrix} 1 & 1 & 5 \\ 1 & 3 & -3 \\ 1 & 5 & -3 \end{vmatrix}; \qquad R3-R2 \qquad \Delta_3 = \begin{vmatrix} 1 & 1 & 5 \\ 1 & 3 & -3 \\ 0 & 2 & 0 \end{vmatrix} = -2\begin{vmatrix} 1 & 5 \\ 1 & -3 \end{vmatrix} = 16.$$

Hence,

$$a_0 = \frac{\Delta_1}{\Delta} = 12; \qquad a_1 = \frac{\Delta_2}{\Delta} = -8; \qquad a_2 = \frac{\Delta_3}{\Delta} = 1.$$

Thus the parabola is $y = 12 - 8x + x^2$.
When $y = 0$, we have $x^2 - 8x + 12 = 0$, that is,

$$(x-2)(x-6) = 0,$$

so $x = 2$ or $x = 6$ make $y = 0$.

Example 9(g) What equation of the form $y = a + bx + cx^2 + dx^3$ connects $(0, -1)$, $(1, 1)$, $(2, 0)$ and $(3, 2)$?

Answer: Substituting the value $(0, -1)$ leads immediately to $a = -1$ so that now $bx + cx^2 + dx^3 = y + 1$. Hence,

with (1, 1) we have $b + c + d = 2$
with (2, 0) we have $2b + 4c + 8d = 1$
with (3, 2) we have $3b + 9c + 27d = 3$.

Then

$$\Delta = \begin{vmatrix} 1 & 1 & 1 \\ 2 & 4 & 8 \\ 3 & 9 & 27 \end{vmatrix} = 12 \qquad \text{(details of calculation omitted.)}$$

$$\Delta_1 = \begin{vmatrix} 2 & 1 & 1 \\ 1 & 4 & 8 \\ 3 & 9 & 27 \end{vmatrix} = 66$$

$$\Delta_2 = \begin{vmatrix} 1 & 2 & 1 \\ 2 & 1 & 8 \\ 3 & 3 & 27 \end{vmatrix} = -54$$

$$\Delta_3=\begin{vmatrix}1 & 1 & 2\\ 2 & 4 & 1\\ 3 & 9 & 3\end{vmatrix}=12$$

$$\therefore\ b=\frac{\Delta_1}{\Delta}=\frac{11}{2};\qquad c=\frac{\Delta_2}{\Delta}=-\frac{9}{2};\qquad d=\frac{\Delta_3}{\Delta}=1.$$

So $2y=2x^3-9x^2+11x-2$.

EXERCISE 9C

Where

$$\Delta=\begin{vmatrix}1 & x_1 & x_1^2 & x_1^3\\ 1 & x_2 & x_2^2 & x_2^3\\ 1 & x_3 & x_3^2 & x_3^3\\ 1 & x_4 & x_4^2 & x_4^3\end{vmatrix}$$

show that $\Delta=(x_2-x_1)(x_3-x_2)(x_3-x_1)(x_4-x_3)(x_4-x_2)(x_4-x_1)$. Find the equation of the form $y=a_0+a_1x+a_2x^2+a_3x^3$ which contains the points $(-2,-12)$, $(-1,0)$, $(1,0)$, and $(2,0)$. Sketch the curve and find the values of x and y at which y is locally a maximum (top of peak) and locally a minimum (bottom of trough).

9.5 THE MATRIX METHOD OF SOLVING SIMULTANEOUS EQUATIONS

A square matrix of order 3×3 operating on a suitable column vector, i.e. a matrix of order 3×1, will produce another column vector similar to the first. Thus matrix

$$\mathbf{M}=\begin{pmatrix}a_1 & a_2 & a_3\\ b_1 & b_2 & b_3\\ c_1 & c_2 & c_3\end{pmatrix}$$

multiplied into

$$\mathbf{V}=\begin{pmatrix}x\\ y\\ z\end{pmatrix}$$

results in

$$\mathbf{W} = \begin{pmatrix} r_1 \\ r_2 \\ r_3 \end{pmatrix},$$

i.e.

$$\mathbf{M} \cdot \mathbf{V} = \mathbf{W}, \tag{9.7}$$

where

$$\begin{aligned} a_1x + a_2y + a_3z &= r_1 \\ b_1x + b_2y + b_3z &= r_2 \\ c_1x + c_2y + c_3z &= r_3. \end{aligned} \tag{9.8}$$

This is, of course, a set of simultaneous equations of the kind we have been examining up to now in this chapter. Obtaining the set in this way stresses the use of a matrix as a transformation operator. The matrix, operating on the column vector **V** which embodies the initial experimental conditions, results in the column vector **W** which embodies the final observed measurements. The practical purpose of solving a set like that of equation (9.8) is to determine the initial conditions x, y, and z from the known final outcome.

Now suppose there were a matrix **R** such that $\mathbf{R} \cdot \mathbf{M} = 1$. If equation (9.7) is multiplied through by **R**, giving $\mathbf{R} \cdot \mathbf{M} \cdot \mathbf{V} = \mathbf{R} \cdot \mathbf{W}$, then since the l.h.s. now is just **V**, what is left is

$$\mathbf{R} \cdot \mathbf{W} = \mathbf{V}. \tag{9.9}$$

There is, in most circumstances, a matrix **R** with the required property. It is called the reciprocal of **M**, and it provides the operation which transforms the set of final observations, **W**, back into the starting conditions, **V**.

There is no need, for our purposes, to derive and prove all the properties of the reciprocal matrix – it is sufficient to know how to calculate it. So here is the procedure for obtaining **R** in the case of a 3×3 matrix like **M**.

1. Transpose the rows and columns of **M**, i.e. write out the **transpose matrix** $\mathbf{M}'$ in which row i of **M** becomes column i of $\mathbf{M}'$ and column j of **M** becomes row j of $\mathbf{M}'$. Thus in the present case

$$\mathbf{M}' = \begin{pmatrix} a_1 & b_1 & c_1 \\ a_2 & b_2 & c_2 \\ a_3 & b_3 & c_3 \end{pmatrix}.$$

2. Replace each element of $\mathbf{M}'$ by its **cofactor**, defined in the following way. If $|\mathbf{M}|$ is the determinant formed by the elements of the matrix **M**, then

each element in the determinant has a minor, as used by us before in expanding and evaluating determinants [cf. equation (8.9)], and each element has a sign attached to it according to the scheme

$$\begin{vmatrix} + & - & + \\ - & + & - \\ + & - & + \end{vmatrix}.$$

The minor of an element, together with its sign, is called that element's cofactor. If the cofactor of an element is shown by the corresponding capital letter, then the matrix thus formed is called the **adjoint** (or sometimes the **adjugate**) matrix (of **M**, in this case) and is written (for either name) adj **M**, so that we have

$$\text{adj}\,\mathbf{M} = \begin{pmatrix} A_1 & B_1 & C_1 \\ A_2 & B_2 & C_2 \\ A_3 & B_3 & C_3 \end{pmatrix}.$$

3. Finally, divide by $|\mathbf{M}|$ to obtain **R**, the reciprocal or inverse of **M**, i.e. $\mathbf{R} = (\text{adj}\ \mathbf{M})/|\mathbf{M}| = \mathbf{M}^{-1}$.

The reciprocal matrix cannot be formed if $|\mathbf{M}| = 0$.

Finally, however, to repeat a point made before in section 9.1, each problem should be examined carefully for the best method of solution. For the solution of really large sets of equations, as found, for example, in the determination of protein structure by X-ray crystallography, the matrix method of solution is used, but computerized to find the most efficient route to calculating the reciprocal matrix. With smaller sets, it may be much more convenient to use Cramer's rule, or to use successive elimination as described in the first section of this chapter.

Example 9(h) Solve the equations of example 9(d) by use of the reciprocal matrix method.

Answer:

$$\mathbf{M} = \begin{pmatrix} 2 & 1 & 2 \\ 1 & -5 & 1 \\ 4 & 2 & -1 \end{pmatrix}, \quad \mathbf{V} = \begin{pmatrix} x \\ y \\ z \end{pmatrix}, \quad \text{and} \quad \mathbf{W} = \begin{pmatrix} 5 \\ 8 \\ 0 \end{pmatrix}.$$

Therefore the transpose matrix is

$$\mathbf{M}' = \begin{pmatrix} 2 & 1 & 4 \\ 1 & -5 & 2 \\ 2 & 1 & -1 \end{pmatrix}.$$

The cofactors of the elements in the first row of $\mathbf{M}'$ are

$$+\begin{vmatrix} -5 & 2 \\ 1 & -1 \end{vmatrix} = +3; \qquad -\begin{vmatrix} 1 & 2 \\ 2 & -1 \end{vmatrix} = -(-5); \qquad +\begin{vmatrix} 1 & -5 \\ 2 & 1 \end{vmatrix} = +11$$

and similarly for the remaining cofactors. Hence

$$\text{adj}\,\mathbf{M} = \begin{pmatrix} 3 & 5 & 11 \\ 5 & -10 & 0 \\ 22 & 0 & -11 \end{pmatrix}.$$

Also $|\mathbf{M}| = 55$ (i.e. $|\mathbf{M}| = \Delta$ of example 9(d)). Now, from example 9(d),

$$\mathbf{W} = \begin{pmatrix} 5 \\ 8 \\ 0 \end{pmatrix} \text{ and from equation (9.9) } \mathbf{V} = \begin{pmatrix} x \\ y \\ z \end{pmatrix} = \mathbf{R} \cdot \mathbf{W}.$$

But $(\text{adj}\,\mathbf{M}) \cdot \mathbf{W} = \begin{pmatrix} 55 \\ -55 \\ 110 \end{pmatrix}$, so dividing by $|\mathbf{M}|$, $\mathbf{R} \cdot \mathbf{W} = \begin{pmatrix} 1 \\ -1 \\ 2 \end{pmatrix} = \mathbf{V}$. Thus the solution of the equations is $x = 1$, $y = -1$, $z = 2$.

EXERCISE 9D

Solve the set of equations in exercise 9B, number 1, by the reciprocal matrix method.

There will be time to audit
The accounts later, there will be sunlight later
And the equation will come out at last.

Louis MacNeice, *Autumn Journal*

10

Differentiation and integration

10.1 THE DERIVATIVES OF A FUNCTION

If there is a functional relationship $y=f(x)$ between the independent variable x and the dependent variable y, then the derivative of y (sometimes called the differential coefficient of y) is a measure of the rate at which y changes when there is a change in x. The procedure of obtaining the derivative is referred to as the differentiation of y with respect to x. Differentiation, together with the inverse procedure of integration, is often called 'the calculus' or, more formally, 'the differential and integral calculus'. You have probably met it before now, but whether you have or not, I hope you don't regard the subject as an arcane mystery. In English-speaking countries Newton is regarded as the inventor of this branch of mathematics, although in German-speaking countries the credit is usually given to Leibnitz. In either case, however, the origin of the calculus is firmly dated to the late seventeenth century so the ideas have now been around for more than three hundred years. You would expect no difficulty in dealing with the biology, chemistry, or physics of the time of Newton and Leibnitz and no more should you find difficulty with their mathematics.

If x increases by a small amount δx to the value $x+\delta x$, then y will increase by an amount δy to $y+\delta y$. The process of differentiating is then the evaluating of δy knowing $f(x)$ and δx, or, more exactly, evaluating the ratio of δy to δx. (NB: the symbol δ is here being used to mean 'a very small amount of' x or y as the case may be. Note that δ is not a factor multiplying x; it is an operation, like Σ or f where f stands for a function (cf. Section 3.1).) In Fig. 10.1, where the distances have been exaggerated for clarity, suppose that F is the point (x, y) and G is the point $(x+\delta x, y+\delta y)$. Then $FH=(x+\delta x)-x$ and $GH=(y+\delta y)-y$ so that

$$\frac{GH}{FH}=\frac{\delta y}{\delta x}=\tan\theta$$

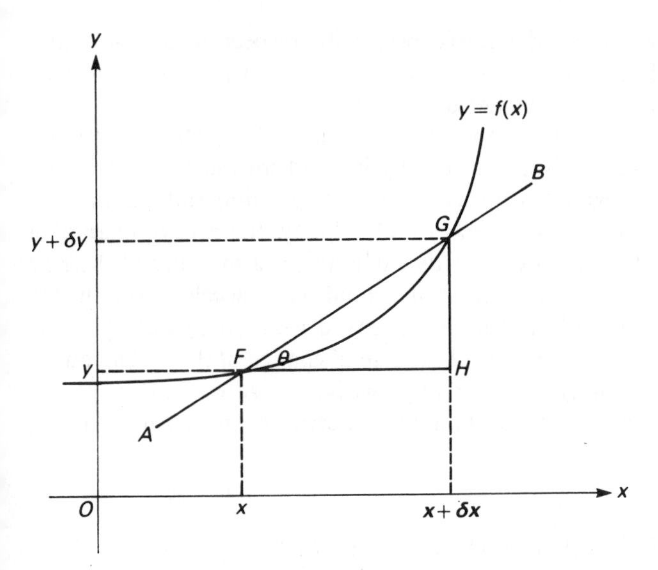

Figure 10.1

where θ is the inclination of the line $AFGB$ to the x-axis. As δx becomes vanishingly small, so also does δy. The point G moves nearer and nearer to F, ending up infinitesimally close to it, and the line $AFGB$, instead of being a chord, becomes a tangent to the curve $y = f(x)$ at the point F. (A chord is a straight line which cuts a curve, whereas a tangent to a curve is defined as a line which just touches the curve, but does not intersect it.) In this situation, $\delta y/\delta x$, which is now written

$$\frac{\mathrm{d}y}{\mathrm{d}x}$$

to show that the limiting (tangential) condition has been reached, is the gradient of the tangent. The quantity $\mathrm{d}y/\mathrm{d}x$ is called the derivative of y with respect to x, or sometimes the first derivative. If x changes by an amount δx, then y changes by an amount

$$\frac{\mathrm{d}y}{\mathrm{d}x}\delta x.$$

If we put $\mathrm{d}y/\mathrm{d}x = z$, then usually z would also be a function of x which could be graphed. In that case, the procedure just enacted could be repeated to give $\mathrm{d}z/\mathrm{d}x$. But remembering that $z = \mathrm{d}y/\mathrm{d}x$, it is usual to maintain the link between y and x by writing $\mathrm{d}z/\mathrm{d}x$ as

$$\frac{\mathrm{d}^2y}{\mathrm{d}x^2},$$

which is called the second derivative of y with respect to x. And since d^2y/dx^2 is also a function of x, we can go on to find a third derivative, d^3y/dx^3, and then on to still higher derivatives.

The construction of Fig. 10.1 shows how the line $AFGB$, when it becomes a tangent to the curve $y = f(x)$, provides by its gradient the rate at which y changes when x is changed. If y were a measure of position and x a measure of time, then dy/dx would be a measure of velocity, the rate of change of position with time. The second derivative, d^2y/dx^2, is a measure of the rate of change of the rate of change – it would give acceleration in the position–time case. The still higher coefficients, d^ny/dx^n say, would give the rate of change of the rate of change of the rate of change of the ... Of course, these rates of change may be positive, negative, or zero; that is, y may increase, decrease, or remain constant as x increases, and so may dy/dx, d^2y/dx^2, etc.

10.2 THE NOTATION OF DERIVATIVES

The notation of dy/dx for the derivative of y with respect to x is clumsy, inconvenient and potentially misleading, but it is hallowed by centuries of usage and so is perhaps irremovable. The most important feature of it to be noted is that it might have been written (and indeed sometimes is) as

$$\frac{d}{dx}y$$

and the symbol d/dx is another case where an operation is being signalled – in this case the operation of differentiation of $f(x)$ with respect to x when $y = f(x)$. When dy/dx appears in equations or statements it is not to be broken up into bits, such as dy and dx; not is it, or any part of it, available to be divided out (cancelled) as if it were a number or a product of numbers.

The same strictures apply to the symbolism for the higher derivatives. The origin of the use of the symbol d^2y/dx^2 for the second derivative is its meaning as

$$\frac{d}{dx}\left(\frac{d}{dx}y\right).$$

The current usage is not the most obvious or the most logical which could have been arranged. In using the symbol d^ny/dx^n remember that no part of it can be separated off or divided out; and treat it with caution.

However, there are other notations which are in use for derivatives, some of which occasionally appear later in this book. The most frequently encountered is the writing of $f'(x)$ or y' for dy/dx when $y = f(x)$. This is then extended in an obvious way by writing $f''(x)$ or y'' for d^2y/dx^2; $f'''(x)$ or y'''

for d^3y/dx^3 etc. but more than three ticks becomes too cumbersome. Some authors have used D for d/dx and then by extension D^2 for d^2/dx^2, but this notation, though convenient, has not caught on. A quite common notation, just in the case where x represents time, and perhaps has the symbol t, is to represent the derivative by a dot. Thus dy/dt could be written as $\dot{y}$, d^2y/dt^2 as $\ddot{y}$, and so on (but again probably not beyond $\dddot{y}$).

10.3 DERIVATIVES OF STANDARD FORMS

Since you have most probably encountered these ideas before, I do not intend to spend much time on them – merely to indicate the general procedure to be followed, to give two or three examples of the operation of differentiating to obtain the derivative *ab initio*, and then to provide you with a reminder list of the most useful results in Table 10.1.

The operation of differentiating

Where $y=f(x)$, suppose that y increases to $y+\delta y$ when x increases to $x+\delta x$. Therefore, $\delta y=f(x+\delta x)-f(x)$. The derivative is defined by

$$\frac{dy}{dx}=\lim_{\delta x\to 0}\frac{f(x+\delta x)-f(x)}{\delta x}. \qquad (10.1)$$

Table 10.1 Derivatives for $y=f(x)$

$f(x)$	$\frac{dy}{dx}$	$f(x)$	$\frac{dy}{dx}$
constant, k	0	$\sin x$	$\cos x$
x^n	nx^{n-1}	$\cos x$	$-\sin x$
kx^n	knx^{n-1}	$\tan x$	$\sec^2 x$
e^x	e^x	$\sec x$	$\sec x \cdot \tan x$
$\ln x$	$\frac{1}{x}$	$\operatorname{cosec} x$	$-\operatorname{cosec} x \cdot \cot x$
k^x	$k^x \ln k$	$\cot x$	$-\operatorname{cosec}^2 x$
$\log_k x$	$\frac{1}{x} \cdot \ln k$	$\left.\begin{matrix}\sin^{-1}x\\ \cos^{-1}x\end{matrix}\right\}$	$\frac{1}{\sqrt{1-x^2}}$
		$\tan^{-1}x$	$\frac{1}{1+x^2}$

Derivative of x^2

When $f(x)=x^2$, we have to deal with $(x+\delta x)^2$ which equals $x^2+2x(\delta x)+(\delta x)^2$.

$$\therefore \quad \frac{dy}{dx}=\lim_{\delta x\to 0}\frac{x^2+2x(\delta x)+(\delta x)^2-x^2}{\delta x} \\ =\lim_{\delta x\to 0}(2x+\delta x). \tag{10.2}$$

As δx tends to zero, the r.h.s. clearly tends to $2x$, which is the derivative of x^2.

Derivative of x^n

If n is a positive integer, a simple binomial expansion gives

$$(x+\delta x)^n=x^n+nx^{n-1}(\delta x)+\frac{n(n-1)}{1\cdot 2}x^{n-2}(\delta x)^2+\cdots$$

Using the definition of equation (10.1) and following the pattern of equation (10.2) we get for the derivative of $y=x^n$

$$\frac{dy}{dx}=\lim_{\delta x\to 0}\left(nx^{n-1}+\frac{n(n-1)}{1\cdot 2}x^{n-2}(\delta x)+\cdots\right) \tag{10.3}$$

In equation (10.3), the second and all further terms within the bracket contain δx and its powers; and all of them necessarily tend to zero as δx tends to zero. Hence we are left with nx^{n-1} as the derivative of x^n.

If n is not a positive integer, it is necessary to make an expansion by means of the Binomial theorem (as described in Section 3.7) to obtain

$$(x+\delta x)^n=x^n\left(1+\frac{\delta x}{x}\right)^n=x^n\left(1+n\frac{\delta x}{x}+\frac{n(n-1)}{1\cdot 2}\left(\frac{\delta x}{x}\right)^2+\cdots\right).$$

It then follows very readily that the same derivative of x^n results, namely nx^{n-1}, whether n is positive or negative, integer, rational or irrational.

Derivative of a constant

If $y=f(x)=k$ where k is constant, then $f(x+\delta x)=f(x)$ and $dy/dx=0$.

Derivative of sin x

If $y=f(x)=\sin x$, we have to deal with $\sin(x+\delta x)$. Using the formula of Table 4.3, we get

$$\sin(x+\delta x)=\sin x\cos\delta x+\cos x\sin\delta x,$$

but as $\delta x \to 0$, $\cos \delta x \to 1$ and $\sin \delta x \to \delta x$ (cf. Section 4.6). Thus when δx is very small,

$$f(x+\delta x)-f(x)=\sin x \cdot 1+(\cos x)\delta x-\sin x$$
$$=(\cos x)\delta x.$$

Then using the definition of equation (10.1), we get $\cos x$ as the derivative of $\sin x$.

10.4 TABLE OF DERIVATIVES

The procedures just used can be applied to all the functions encountered hitherto in this book, and to all other functions which are continuous – that is, functions which can be graphed as smooth lines not containing sudden jumps or sharp angles. Provided this condition is fulfilled, and also one or two others relating to extreme conditions only rarely encountered, the derivatives can be calculated. But in usual practice, most complicated functions can be dissected and the derivatives can then be assembled from the derivatives of the constituent elementary parts. Table 10.1 lists the most frequently encountered examples of the latter.

10.5 DIFFERENTIATING A MIXTURE OF FUNCTIONS

Formulae for the derivatives of sums, products, quotients and other mixtures of functions can all be obtained by appropriately applying the definition of equation (10.1). However, it is sufficient for you merely to note the results and some accompanying comments. The results are given in Table 10.2 and the comments follow. The notation used is that in which, when $y=f(x)$, y' or $f'(x)$ signifies the derivative of the function $f(x)$.

Table 10.2 Derivatives of mixtures of functions

	y	$\frac{dy}{dx}$
Sum:	$f(x)+g(x)$	$f'(x)+g'(x)$
Product:	$f(x)\cdot g(x)$	$f'(x)\cdot g(x)+f(x)\cdot g'(x)$
Quotient:	$\frac{f(x)}{g(x)}$	$\frac{f'(x)\cdot g(x)-f(x)\cdot g'(x)}{[g(x)]^2}$
Function of a function:	$f(z)$ where $z=g(x)$	$f'(z)\cdot z'=f'(z)\cdot g'(x)$

Sum of functions

The simple result here is what one might expect. Differentiation is distributive across addition of functions, so that the derivative of the sum of functions is the sum of the separate derivatives.

Product of functions

Differentiation is not distributive across a product; it is necessary to differentiate the terms of the product one at a time, leaving it multiplied by the other(s) and then add up the results. If $y=rs$ where r and s are, separately, functions of x, then

$$y'=r's+rs'.$$

This extends to larger products. So if $y=rst$,

$$y'=r'st+rs't+rst'$$

where t is an additional function of x, and so on. In the case where one of the functions is a constant, the derivative of y is just the constant multiplied by the derivative of the remaining part of the product, i.e. if $y=a \cdot f(x)$, then $y'=a \cdot f'(x)$.

Quotient of two functions

It is best just to learn the formula in Table 10.2. I was taught this mnemonic by Mr H. G. Dixon: 'D-numerator, D-nominator minus numerator D-D-nominator over D-nominator squared'. (The upper element in a fraction is its numerator and the lower one its denominator.) Alternatively, the quotient can be considered as the product of two functions, $f(x)$ and $h(z)$ where $h(z)=1/z$ and $z=g(x)$; so $h(z)$ is a function of a function and to differentiate it you must know the next formula on the list.

Function of a function

If $y=f(r)$ and $r=g(s)$ and $s=h(t)$ and ... $=w(x)$, there is a theorem which states that

$$\frac{dy}{dx}=\frac{dy}{dr}\cdot\frac{dr}{ds}\cdot\frac{ds}{dt}\cdots\frac{dw}{dx}.$$

This is what the entry in Table 10.2 says, but with only two functions involved. If $y=f(x)$ where $f(x)$ is either an exponential function or a complicated product it may be worth creating a function of a function by taking logarithms (to base e). Taking logarithms gives

$$\ln y = \ln f(x)$$

and differentiating with respect to x, since

$$\frac{\mathrm{d}}{\mathrm{d}y}(\ln y) = \frac{1}{y},$$

gives

$$\frac{1}{y}\frac{\mathrm{d}y}{\mathrm{d}x} = \frac{\mathrm{d}}{\mathrm{d}x}(\ln f(x))$$

so that

$$\frac{\mathrm{d}y}{\mathrm{d}x} = y \cdot \frac{\mathrm{d}}{\mathrm{d}x}(\ln f(x)).$$

Summary

The rules and formulae given in Tables 10.1 and 10.2 may appear complicated, but that is chiefly because there are so many of them. In practice, they are found to be quite simple and rarely give cause for error, provided you keep a careful watch out for functions of functions. The best way for you to relearn or revise them is, as always, to look carefully at the examples and then to do the exercises.

Example 10(a) Find the derivatives of:

(a) $y_1 = 5x^2 + 3x + 1$;

(b) $y_2 = \sin(2\pi x + \alpha)$, where α is a constant;

(c) $y_3 = \dfrac{1}{\sqrt{1 + x^2}}$;

(d) $y_4 = x^2 \cdot \sin x^2$;

(e) $y_5 = x^x$.

Answer:

(a) If $y = kx^n$,

$$\frac{\mathrm{d}y}{\mathrm{d}x} = knx^{n-1}.$$

$$\therefore \qquad \frac{\mathrm{d}}{\mathrm{d}x}(5x^2) = 10x.$$

Hence

$$\frac{dy_1}{dx} = 10x + 3.$$

(b) Write $y_2 = \sin z$ where $z = 2\pi x + \alpha$ and $dz/dx = 2\pi$. Then

$$\frac{dy_2}{dx} = \frac{dy_2}{dz} \cdot \frac{dz}{dx} = \cos(z) \cdot 2\pi$$

$$= 2\pi \cos(2\pi x + \alpha).$$

(c) Write $y_3 = 1/r$ where $r = s^{\frac{1}{2}}$ and $s = 1 + x^2$. Then

$$\frac{dy_3}{dx} = \frac{dy_3}{dr} \cdot \frac{dr}{ds} \cdot \frac{ds}{dx}.$$

But

$$y_3 = r^{-1} \qquad \text{so that } \frac{dy_3}{dr} = -1 \cdot r^{-2}$$

$$r = s^{\frac{1}{2}} \qquad \text{so that } \frac{dr}{ds} = \tfrac{1}{2} s^{-\frac{1}{2}}$$

$$s = 1 + x^2 \qquad \text{so that } \frac{ds}{dx} = 2x.$$

So

$$\frac{dy_3}{dx} = -\frac{1}{r^2} \cdot \frac{1}{2s^{\frac{1}{2}}} \cdot 2x$$

$$= -x(1 + x^2)^{-\frac{3}{2}}.$$

(d) Put $y_4 = uv$ where $u = x^2$ and $v = \sin x^2$. Put $v = \sin z$ where $z = x^2$. Then

$$v' = \cos(z) \cdot 2x = 2x \cdot \cos(x^2).$$

Now, $y_4' = u'v + uv'$ and $u' = 2x$.

$$\therefore \qquad \frac{dy_4}{dx} = 2x \cdot \sin x^2 + 2x^3 \cdot \cos x^2.$$

(e) Take logarithms to base e so that $\ln y_5 = x \ln x$. Differentiate with respect to x:

$$\frac{1}{y_5} \frac{dy_5}{dx} = \ln x + x\left(\frac{1}{x}\right).$$

Thus

$$\frac{dy_5}{dx} = x^x(\ln x + 1).$$

EXERCISES 10A

1. Find the derivatives of the following functions:
   (a) $6x^3+4x^2+2x+1$;
   (b) x^2+1+x^{-2};
   (c) $\tan(\pi x^2/2)$;
   (d) $(1+2x+3x^2)^{\frac{3}{2}}$.
2. Find the derivatives of the following functions:
   (a) $e^{-x}\cos(x^2)$;
   (b) $\dfrac{1}{1+x^2}$;
   (c) $\sqrt{\dfrac{1-x}{1+x}}$;
   (d) $a^{x\sin x}$.
3. The BMR (basal metabolic rate) B, for vertebrates is given approximately by $B=1.78W^{0.73}$ where W is the body weight in kilograms. For what body weight would the BMR increase twice as fast as body weight? (Hint: the answer is the value of W for which $dB/dW=2$.)

10.6 MAXIMA AND MINIMA

If the graph of the function $y=f(x)$ is a curve with peaks and troughs, like those of Figs 7.4 and 7.5 or Fig. 10.2, then as x takes on values which approach a peak from the left hand side, i.e. as x increases, y will also increase, but will do so at a decreasing rate such that just at the peak the rate of increase of y is zero. Thereafter it begins to fall so that its 'rate of increase' is negative, giving in fact a rate of decrease. But the rate of change

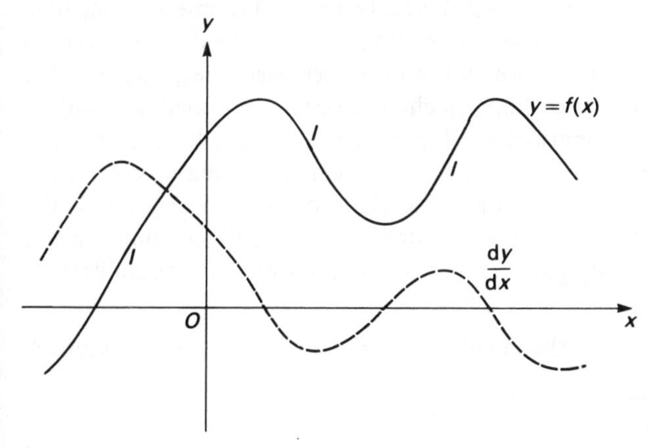

Figure 10.2

– increase or decrease – is exactly what is measured by dy/dx. So as y ascends a peak (going from left to right), reaches it, and then descends on the other side, dy/dx goes from positive values through zero to negative values. By an exactly similar line of argument, as y descends into a trough, reaches the bottom of it, and then begins to ascend on the other side, dy/dx starts negative but rises to zero and then to positive values. Thus, where there is a functional relationship between x and y, values (x, y) for which $dy/dx=0$ are either maxima or minima of y.

If you find the x and y values which make $dy/dx=0$, how are you to tell, without drawing the graph, whether the point you have obtained is at a maximum or a minimum position? Look back at what dy/dx was doing in the two situations. As y was approaching, reaching, and then proceeding beyond a maximum, dy/dx was falling, but if, as x increases, dy/dx is falling, its rate of change is negative. The rate of change of dy/dx is measured as d^2y/dx^2. Thus if d^2y/dx^2 is negative with the values of x and y for which $dy/dx=0$, the point (x, y) is at a maximum. By an exactly corresponding argument, if d^2y/dx^2 were found to be positive, the point would be at a minimum.

If the second derivative, d^2y/dx^2, is sometimes positive and sometimes negative, then between these conditions there are values of (x, y) at which it is zero. What happens there?

In just the same way that $dy/dx=0$ shows where y is a maximum or minimum, so $d^2y/dx^2=0$ shows where dy/dx is a maximum or minimum. But what is the meaning of a maximum or minimum value of dy/dx? Take the case of a minimum and suppose it occurs in conditions where dy/dx is negative, as in Fig. 10.2. Negative dy/dx means that y is decreasing as x increases. But as dy/dx approaches, reaches and passes through a minimum, with y all the while decreasing, the rate of its decrease slows down, stops (ceases to move to larger magnitudes), and begins to become less negative (starts to fall in magnitude). This allows y to continue to decrease, but to switch over from decreasing more and more to decreasing less and less. On a graph, the point at which such a change occurs is called a point of inflexion: such points are marked I in Fig. 10.2. In everyday life the situation is met, for example, in driving a car, when the car is slowing down but the driver switches over from braking to accelerating and the sound of the engine revolutions signals that, even though the car is still slowing down, it is about to pick up speed again. Table 10.3 summarizes the possibilities.

Example 10(b) Determine the maxima, minima, and points of inflexion of:

(a) $y=x^3+4x^2-3x+7$;
(b) $y=(x^2-1)(x^2-4)$.

Table 10.3 Conditions for maxima, minima and inflexions

$\frac{dy}{dx}$	$\frac{d^2y}{dx^2}$	y
0	negative	at a maximum
0	positive	at a minimum
maximum or minimum	0	point of inflexion

Answer:

(a) The function is similar to that graphed in Fig. 7.4.

$$y' = 3x^2 + 8x - 3 \qquad \text{and} \qquad y'' = 6x + 8.$$

The maximum and minimum occur when $y' = 0$, i.e. when $3x^2 + 8x - 3 = 0$ or

$$(3x - 1)(x + 3) = 0$$

or when $x = -3$ or $1/3$.

When $x = -3$, $y'' = -10$ and since it is negative, there is a maximum at that point, where $y = 25$.

When $x = 1/3$, $y'' = +10$ and since it is positive, there is a minimum at that point, where $y = 6.48$.

The point of inflexion comes between the maximum and the minimum, where $y'' = 0$, so that $x = -4/3$ and then $y = 15.74$.

(b) The function is similar to that graphed in Fig. 7.5.

$$\begin{aligned} y' &= 2x(x^2 - 4) + 2x(x^2 - 1) \\ &= 2x(2x^2 - 5) \\ y'' &= 2(2x^2 - 5) + 2x \cdot 4x \\ &= 12x^2 - 10. \end{aligned}$$

The maximum and minima occur when $y' = 0$, i.e. when $2x(2x^2 - 5) = 0$ so that $x = 0$ or $\pm\sqrt{5/2}$.

When $x = 0$, $y = 0$, $y'' = -10$, i.e. y'' is negative, so the point is a maximum.

When $x = +\sqrt{5/2}$ or $-\sqrt{5/2}$, $y = 2.25$, and $y'' = +20$, i.e. y'' is positive so both points are minima.

Points of inflexion are found by putting $y'' = 0$; this gives $x = \pm\sqrt{5/6}$. For both $x = +\sqrt{5/6}$ and $x = -\sqrt{5/6}$, $y = 0.53$.

Example 10(c) Loaves are to be baked of a fixed weight and also of a stated length to fit on to supermarket trays. Hence to keep the volume constant the cross-section is of a fixed area, A, but its shape can be varied. If the shape is that of a rectangle of width $2a$ and height $2b$ surmounted by a semicircle of radius a, as shown in Fig. 10.3, find the values of a and b so that there is minimum crust on the bread (to minimize its rate of drying out).

Answer: Area $A=2a\cdot 2b+\frac{1}{2}(\pi a^2)=4ab+\pi a^2/2$, and A is constant. The length of crust $C=4b+2a+\pi a$.
But

$$b=\frac{A-\pi a^2/2}{4a}=\frac{A}{4a}-\frac{\pi a}{8}.$$

$$\therefore \qquad C=(2+\pi)a+\frac{A}{a}-\frac{\pi a}{2}$$

$$=\left(2+\frac{\pi}{2}\right)a+\frac{A}{a}.$$

(The graph $Ca=(2+\pi/2)a^2+A$ is a hyperbola as shown in Fig. 10.3.) Hence

$$\frac{\mathrm{d}C}{\mathrm{d}a}=2+\frac{\pi}{2}-\frac{A}{a^2}$$

which is zero when $2+\pi/2=A/a^2$ i.e. when

$$a^2=\frac{2A}{4+\pi}.$$

In taking the square root, it is only the positive square root which is of

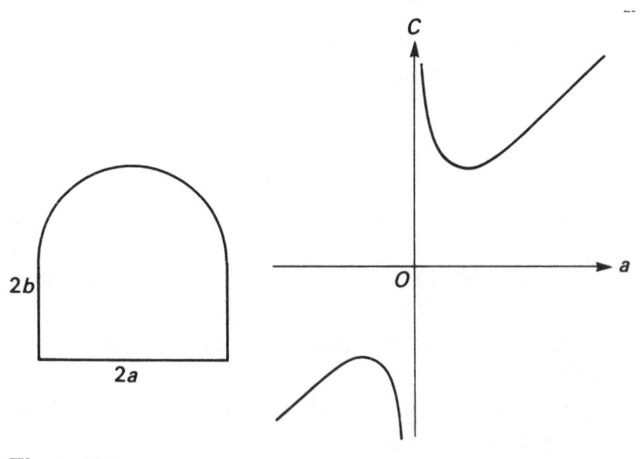

Figure 10.3

interest since only in Alice's world Through the Looking Glass could loaves of bread have negative lengths of crust. Hence $a=\sqrt{2A/(4+\pi)}$ for which d^2C/da^2 is positive, so that the value of C is then a minimum. The value for b is $\sqrt{A/(2(4+\pi))}$ and the ratio $a{:}b$ is 2:1.

EXERCISES 10B

1. Find the positions of the maxima and minima (stating which is which) for the functions:
   (a) $y=5x^3-3x^2-6x+14$,
   (b) $y=2x^4+9x^3-x^2-7$.
2. A farmer has 40 m of movable fencing with which to make a pen to hold as many sheep as possible.
   (a) If the pen is rectangular, show that the maximum area he can enclose is $100\,\text{m}^2$.
   (b) How does this compare with the area of the (approximately) circular pen he could construct?
   (c) What would be the maximum area enclosed if the farmer used a straight river as one side of the enclosure and used the fencing to make the other three sides of a rectangle?
3. A transistor in a radio circuit has an amplification factor proportional to $x/(x^2+a^2)$ where x is a variable and a is a constant. Show that the amplification is a maximum when $x=a$. Is there a value of x for which the amplification is a minimum?

10.7 PARTIAL DIFFERENTIATION

I have been considering so far in this account the way in which a function $f(x)$ changes if there is an increase in the independent variable, x, but I have mentioned before, in chapters 6 and 9, the possibility of a function depending on several variables. So the question arises of whether one can calculate the increase of a variable which is a function dependent on more than one other variable. If is sufficient to limit the description of the situation to the case in which a dependent variable, z, is a function of only two independent variables, x and y. The relationship is written as $z=f(x, y)$. This immediately reveals the problem, because the equation defining z would be graphed as a surface in three dimensions as shown in Fig. 10.4; and in general from any point on the surface one could proceed in an infinity of directions, some of which might lead to an increase in z, some to a decrease, and some to neither.

Whether z increases, decreases, or remains stationary depends on whether x and y are separately incremented or decremented, and so the change in z is calculated with the changes in x and y taken apart, which is why the

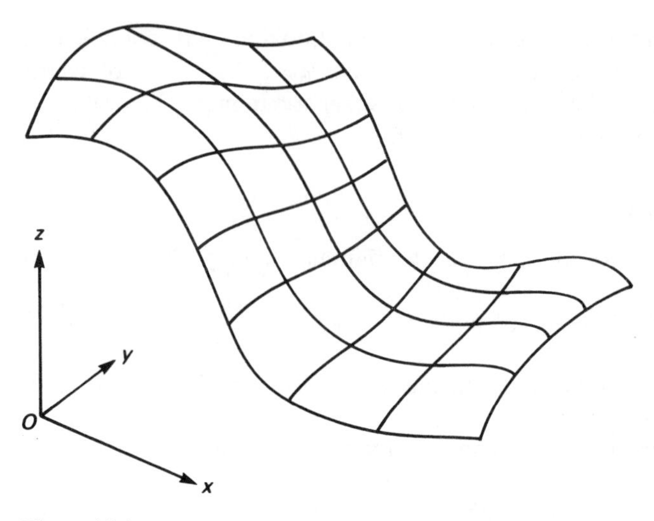

Figure 10.4

procedure is called partial differentiation. What is called the partial derivative of z with respect to x, written as

$$\frac{\partial z}{\partial x},$$

is the derivative of z with respect to x while y is held constant; and similarly,

$$\frac{\partial z}{\partial y}$$

is the derivative of z with respect to y while x is held constant. The total change in z after changes δx and δy in x and y is then given by δz where

$$\delta z = \frac{\partial z}{\partial x}\delta x + \frac{\partial z}{\partial y}\delta y. \tag{10.4}$$

If it were the case that x and y changed in a known way dependent on another variable, δz could be specified more precisely. Specifically, if x and y depended on time, t, in such a way that $\mathrm{d}x/\mathrm{d}t$ and $\mathrm{d}y/\mathrm{d}t$ were known, then dividing equation (10.4) by δt and proceeding to the limit of infinitesimally small δx, δy, and δt

$$\frac{\mathrm{d}z}{\mathrm{d}t} = \frac{\partial z}{\partial x}\frac{\mathrm{d}x}{\mathrm{d}t} + \frac{\partial z}{\partial y}\frac{\mathrm{d}y}{\mathrm{d}t}.$$

Functions of two variables (or more), like functions of a single variable, typically have second and higher derivatives as well as first derivatives. An alternative notation for the first derivatives is z_x for $\partial z/\partial x$ and z_y for $\partial z/\partial y$. Following the notation used for functions of a single variable, $\partial^2 z/\partial x^2$ is the second derivative of z with respect to x, also written sometimes as z_{xx}, and the second derivative of z with respect to y is written similarly as $\partial^2 z/\partial y^2$ or z_{yy}. The second derivative $\partial^2 z/\partial x^2$ is

$$\frac{\partial}{\partial x}(z_x)$$

when y is still held constant; and in the same way, $\partial^2 z/\partial y^2$ can be obtained while x is still held constant. However, in general, if $z = f(x, y)$, then z_x will be a function of both x and y and it would be possible to obtain the derivative of z_x with respect to y; this derivative is written as $\partial^2 z/\partial y \partial x$ or z_{xy}. Alternatively, one could obtain the derivative with respect to x of z_y, giving the second derivative $\partial^2 z/\partial x \partial y$ or z_{yx}. Fortunately, however, it can be shown that $\partial^2 z/\partial y \partial x = \partial^2 z/\partial x \partial y$ or $z_{xy} = z_{yx}$.

Hence, summarizing the procedure of partially differentiating z when $z = f(x, y)$:

1. there are two first partial derivatives, $\partial z/\partial x$ (or z_x) and $\partial z/\partial y$ (or z_y), where each is obtained by the standard rules of differentiation while the remaining independent variable is held constant;
2. there are three second partial derivatives $\partial^2 z/\partial x^2$ (or z_{xx}), $\partial^2 z/\partial y^2$ (or z_{yy}), and $\partial^2 z/\partial x \partial y$ (or z_{xy} which is equal to $\partial^2 z/\partial y \partial z$ or z_{yx}) obtained also by holding the appropriate variable constant while the differentiation proceeds;
3. the system continues in the same way for still higher derivatives and also for the derivatives of a function of three or more independent variables.

Example 10(d) Find the first and second partial derivatives of z where $z = e^{x^2 y^2}$.

Answer: Since

$$\frac{\partial}{\partial x}(x^2 y^2) = 2xy^2,$$

we have

$$\frac{\partial z}{\partial x} = 2xy^2\, e^{x^2 y^2}.$$

Similarly,

$$\frac{\partial z}{\partial y} = 2x^2 y\, e^{x^2 y^2}.$$

Then

$$\frac{\partial^2 z}{\partial x^2} = 2y^2 \, e^{x^2y^2} + 4x^2y^4 \, e^{x^2y^2} = 2y^2(1 + 2x^2y^2) \, e^{x^2y^2}$$

$$\frac{\partial^2 z}{\partial y^2} = 2x^2 \, e^{x^2y^2} + 4x^4y^2 \, e^{x^2y^2} = 2x^2(1 + 2x^2y^2) \, e^{x^2y^2}$$

and

$$\begin{aligned}\frac{\partial^2 z}{\partial x \partial y} &= 4xy \, e^{x^2y^2} + 2xy^2 \cdot 2x^2y \cdot e^{x^2y^2} \\ &= 4xy(1 + x^2y^2) \, e^{x^2y^2}.\end{aligned}$$

EXERCISE 10C

Find the first and second partial derivatives of

$$z = \frac{x^2y^2}{x^2 + y^2}$$

and prove that $\partial^2 z/\partial x \partial y = \partial^2 z/\partial y \partial x$.

10.8 INTEGRATION

Mathematicians have several ways of considering the procedure of integration, but it is simplest for most purposes to regard it just as the inverse of differentiation: there is a functional relationship $y = f(x)$ which you seek, knowing only $\mathrm{d}y/\mathrm{d}x = f'(x)$. As with differentiation, much integration is done on standard elementary functions. Many of the integrals of these functions can be inferred from the list of derivatives in Table 10.1, but for convenience they are re-expressed specifically as integrals in Table 10.4. However, there remain a few points which need special consideration, particularly the way of integrating functions of functions and products of functions.

The notation of integration is the well-known symbol $\int \ldots \mathrm{d}x$. It is best to regard this as a single symbol, something like a bracket, and not to be separated into its constituent parts, $\int$ and $\mathrm{d}x$. It is read as a unity – 'the integral of ... with respect to x'. The origin of the symbolism is the use of S distorted into $\int$ to indicate the sum of infinitesimally small quantities, like δy in the case of differentiation, taken in the limit, when $\delta x \to 0$ and

$$\frac{\delta y}{\delta x} \to \frac{\mathrm{d}y}{\mathrm{d}x}.$$

Although again, like the case of differentiation, the symbolism is somewhat

Table 10.4 Integrals $I=\int y\,dx$ (but constant omitted)

y	I		y	I
constant, k	kx		$\sin x$	$-\cos x$
x^n	$\frac{x^{n+1}}{n+1}$	$n\neq -1$	$\cos x$	$\sin x$
kx^n	$\frac{kx^{n+1}}{n+1}$			
e^x	e^x			
$\ln x$	$x\ln x+x$		$\frac{1}{\sqrt{1-x^2}}$	$\sin^{-1}x$ or $\cos^{-1}x$
k^x	$k^x/\ln k$			
$\frac{1}{x}$	$\ln x$		$\frac{1}{1+x^2}$	$\tan^{-1}x$

inconvenient, you are very unlikely to meet any alternative symbolism, so there is no occasion for notational confusion in the procedure of integration.

10.9 INDEFINITE AND DEFINITE INTEGRALS

Consider the three functions $y=x^2-4$, $y=x^2$ and $y=x^2+4$. Each graphs as a parabola, of which the first cuts the x-axis at $x=+2$ and $x=-2$, when $y=0$, the second has the x-axis as a tangent at $x=0$, and the third is always quite distant from the x-axis (cf. section 7.4), so the three parabolas have distinctly different properties. Yet in each case, the derivative of y with respect to x is the same. Given only that $dy/dx=2x$, integration will not reveal which parabola is being considered. The procedure of integration can reverse the procedure of differentiation only to within a constant value and that value can be found only if further information is provided. An integral of this kind, whether or not the further information is forthcoming, is called an indefinite integral. If $dy/dx=f'(x)$, integration gives

$$y=\int f'(x)\,dx=f(x)+C,$$

where C, called the constant of integration, is to be determined by further information, if available.

An easier situation arises if what is required is just the *change* in y as x

increases from, say, x_1 to x_2. If the corresponding values of y are y_1 and y_2, then we would have

$$y_1 = f(x_1) + C$$
$$y_2 = f(x_2) + C.$$

Hence, the change in y is

$$y_2 - y_1 = f(x_2) - f(x_1).$$

To find this quantity, the value of the constant of integration is irrelevant. In this case the calculation proceeds by way of what is called a definite integral, written

$$\int_{x_1}^{x_2} f'(x)\,\mathrm{d}x$$

and read as 'the integral of ... with respect to x from x_1 to x_2'.

The operation of integration is often introduced in textbooks by way of definite integrals applied to finding the area between a curve $y = F(x)$ and the x-axis, bounded by vertical lines at x_1 and x_2. The area, A, is the sum of the areas of thin strips like $ABCD$ in Fig. 10.5. If D is at x and C at $x + \delta x$, the width of the strip is δx, and if $AD = y$ and $BC = y + \delta y$, then the area of the strip is $y\delta x$ plus a little bit (the tiny triangle at the top of the strip of area approximately $\frac{1}{2}\delta x \delta y$) which is so small that it can be ignored. So as x increases by δx, we can suppose that A increases by δA and $\delta A = y\delta x$. In the

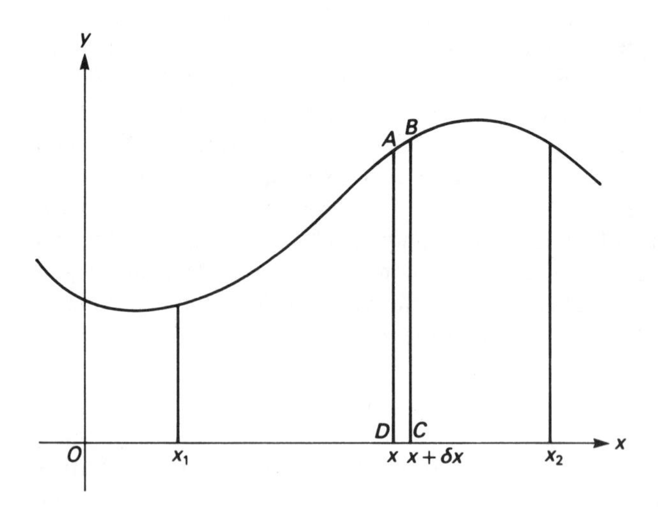

Figure 10.5

limit,

$$\frac{dA}{dx} = y \qquad \text{and} \qquad A = \int_{x_1}^{x_2} y\,dx,$$

where the limits x_1 and x_2 are included to show the confines of the region of interest. If we suppose in this example that $F(x) = f'(x)$ where, as before, $f'(x)$ is the derivative of $f(x)$, then

$$A = \int_{x_1}^{x_2} F(x)\,dx = f(x_2) - f(x_1).$$

It is usual to write interim lines of derivation with the notation

$$\int_{x_1}^{x_2} f'(x)\,dx = [f(x)]_{x_1}^{x_2}$$
$$= f(x_2) - f(x_1);$$

the superscript and subscript to the big square brackets then serve as reminders of the limits of integration which are to be entered into the calculation.

Example 10(e) What are the (indefinite) integrals of:

(a) $5x^4$;
(b) x^3;
(c) $3x^2 + 2x + 1$;
(d) $\cos x$;
(e) $\sin x$?

Answer: In each case, go for the obvious answer, as the inverse of differentiation, but (i) don't forget the constant of integration, and (ii) mentally check the answer by differentiating and hence returning to the original expression.

(a) $x^5 + C$;
(b) $\frac{x^4}{4} + C$;
(c) $x^3 + x^2 + x + C$;
(d) $\sin x + C$;
(e) $C - \cos x$.

Example 10(f) Show that

$$\frac{1}{2}\int_0^{\pi} \sin x\,dx = \int_0^{\pi/4} \cos 2\theta\,d\theta.$$

Answer:

$$\frac{1}{2}\int_0^{\pi/2} \sin x\,\mathrm{d}x = \tfrac{1}{2}[-\cos x]_0^{\pi/2} = \tfrac{1}{2}(-0-(-1)) = \tfrac{1}{2}.$$

$$\int_0^{\pi/4} \cos 2\theta\,\mathrm{d}\theta = [\tfrac{1}{2}\sin 2\theta]_0^{\pi/4} = \tfrac{1}{2}(1-0) = \tfrac{1}{2}.$$

Example 10(g) Find the area under the curve $y=(x+1)^{-2}$ between $x=0$ and $x=2$.

Answer: $(x+1)^{-2}$ can be recognized as

$$-\frac{\mathrm{d}}{\mathrm{d}x}(x+1)^{-1}.$$

Hence

$$A=\int_0^2 y\,\mathrm{d}x=\int_0^2 \frac{\mathrm{d}x}{(x+1)^2}=-\left[\frac{1}{x+1}\right]_0^2=-\frac{1}{2+1}+\frac{1}{0+1}=\frac{2}{3}.$$

EXERCISES 10D

1. Find the (indefinite) integrals of
   (a) $10x^9+7x^6-4x^3$;
   (b) $2x^3+x$;
   (c) $\sec\alpha x\cdot\tan\alpha x$;
   (d) $\sin\frac{x}{3}\cdot\cos\frac{x}{3}$ (Hint: use the formula for $\sin 2A$);
   (e) $(1+x)^{-1}$;
   (f) $(1+x^2)^{-1}$.
2. Find the area under the curve $y=4x^4-x+7$ between $x=0$ and $x=1$.

10.10 TECHNIQUES OF INTEGRATION

Changing the variable

It is most important, in setting out to integrate a function, to examine closely its structure. If what is given as $I=\int f(x)\,\mathrm{d}x$ can be considered as $I=\int h'[g(x)]g'(x)\,\mathrm{d}x$ then $I=h[g(x)]+C$. The implicit procedure here is the substitution $g(x)=z$ so that $I=h(z)+C$ and $\mathrm{d}I/\mathrm{d}x=h'(z)g'(x)$. Explicitly, substitute z for $g(x)$ under the integral sign, and $\mathrm{d}z$ for $g'(x)\,\mathrm{d}x$, so that the problem is presented as $I=\int h(z)\,\mathrm{d}z$.

Example 10(h) Evaluate

(a) $\displaystyle\int \frac{2x}{x^2+1}\,dx$,

(b) $\displaystyle\int \tan x\,dx$.

Answer:
(a) Since

$$\frac{d}{dx}(x^2+1)=2x,$$

put $z=x^2+1$ and dz in place of $2x\,dx$.
Then

$$I=\int\frac{dz}{z}=\ln z+C=\ln(x^2+1)+C.$$

(b) $\tan x=\sin x/\cos x$ and

$$\sin x=-\frac{d}{dx}(\cos x).$$

Hence put $z=\cos x$ and dz in place of $-\sin x\,dx$. The

$$I=-\int\frac{dz}{z}=C-\ln z=C-\ln(\cos x).$$

(NB: This assumes that $\cos x$ is positive; $\ln z$ has no meaning for $z<0$.)

Integration by parts

Corresponding to finding the derivative of the product of two functions there is the problem of finding the integral of such a product; however, the result is always called integration by parts. Suppose that U and V are functions of x for which $dU/dx=u$ and $dV/dx=v$, where u and v are also functions of x. Then $d/dx\,(UV)=uV+Uv$ so that

$$uV=\frac{d}{dx}(UV)-Uv.$$

Integrating both sides of this equation gives

$$\begin{aligned}\int uV\,dx&=\int\frac{d}{dx}(UV)\,dx-\int Uv\,dx\\&=UV-\int Uv\,dx.\qquad(10.5)\end{aligned}$$

But $U=\int u\,dx$, so equation (10.5) states (in the rule to be remembered for integration by parts):

The integral of the product of two functions (u and V) equals the integral of the first (U) times the second (V) minus the integral of [the integral of the first (U) times D of the second (v)].

Here it is again, in different symbols:

$$\int f(x)g(x)\,dx = \left[\int f(x)\,dx\right]g(x) - \int\left[\int f(x)\,dx\right]\frac{d}{dx}g(x)\,dx.$$

Note that if the initial integral is indefinite, then a constant of integration must be added after evaluation of the outer integral in the second term on the r.h.s. If the initial integral is definite, then the upper and lower limits apply to all of the r.h.s. – to the first term as it stands, above – and to the second by making the outer integral a definite one, between the limits applying on the l.h.s.

Example 10(i) Integrate $x\,e^x$.

Answer: Integrate by parts but first putting

$$I = \int x\,e^x\,dx$$

$$= \int e^x \cdot x \cdot dx.$$

The order of the functions in I is switched round because integrating or differentiating e^x gives e^x whereas differentiating x gives 1.

Thus

$$I = e^x \cdot x - \int e^x \cdot 1\,dx$$

$$= x\,e^x - e^x + C.$$

Example 10(j) Integrate $x^2(x+1)^{\frac{1}{2}}$.

Answer: Again integrate by parts after inverting the order of the functions x^2 and $(x+1)^{\frac{1}{2}}$.
Thus

$$I = \int (x+1)^{\frac{1}{2}}x^2\,dx$$

$$= \tfrac{2}{3}(x+1)^{\frac{3}{2}}x^2 - \int \tfrac{2}{3}(x+1)^{\frac{3}{2}}2x\,dx.$$

But

$$\int (x+1)^{\frac{3}{2}} x \, dx = \tfrac{2}{5}(x+1)^{\frac{5}{2}} x - \int \tfrac{2}{5}(x+1)^{\frac{5}{2}} \cdot 1 \, dx$$

$$= \tfrac{2}{5} x (x+1)^{\frac{5}{2}} - \tfrac{4}{35}(x+1)^{\frac{7}{2}} + C.$$

So

$$I = \tfrac{2}{3} x^2 (x+1)^{\frac{3}{2}} - \tfrac{8}{15} x (x+1)^{\frac{5}{2}} + \tfrac{16}{105}(x+1)^{\frac{7}{2}} + C.$$

Example 10(k) Integrate $\ln x$ (cf. Table 10.4).

Answer: Regard $\ln x$ as $1 \cdot \ln x$, where 1 is the first function. Then

$$I = \int 1 \cdot \ln x \, dx$$

$$= x \cdot \ln x - \int x \cdot \frac{1}{x} \, dx \qquad \left(\text{since } \frac{d}{dx}(\ln x) = \frac{1}{x}\right)$$

$$= x \cdot \ln x - x + C.$$

Example 10(1) Evaluate $I = \int e^x \sin x \, dx$.

Answer: To find I it is convenient to consider at the same time

$$J = \int e^x \cos x \, dx.$$

Integrating by parts,

$$I = e^x \sin x - \int e^x \cos x \, dx$$

$$= e^x \sin x - J,$$

so that

$$J + I = e^x \sin x. \tag{10.6}$$

But similarly,

$$J = e^x \cos x - \int e^x(-\sin x) \, dx$$

$$= e^x \cos x + I$$

so that

$$J - I = e^x \cos x. \tag{10.7}$$

Subtracting equation (10.7) from equation (10.6),

$$2I = e^x \sin x - e^x \cos x$$

Hence

$$I = \tfrac{1}{2}(e^x \sin x - e^x \cos x).$$

Note that J is obtained from adding equations (10.6) and (10.7), giving

$$J = \tfrac{1}{2}(e^x \sin x + e^x \cos x).$$

Integration by substitution

This important technique is the integration procedure which inverts finding the derivative of a function of a function. If the variable to be integrated is $y = f(x)$, and if f is suitable, then the procedure is to substitute $x = g(z)$ and it is not difficult to show that

$$\int f(x)\,dx = \int f[g(z)]g'(z)\,dz$$

where $dx/dz = g'(z)$. Unfortunately, 'suitable' in this context is a slippery idea: it means that you need insight to spot, or must find by trial and error, what substitution to make and must also check whether it actually helps to evaluate the integral.

Example 10(m) Evaluate

$$I = \int \frac{dx}{(a^2 - x^2)^{\frac{3}{2}}}.$$

Answer: Put $x = a \sin \theta$ so that $dx/d\theta = a \cos \theta$.
Hence

$$I = \int \frac{a \cos \theta \, d\theta}{(a^2 - a^2 \sin^2 \theta)^{\frac{3}{2}}}$$

$$= \int \frac{d\theta}{a^2 \cos^2 \theta}$$

since $a^2 - a^2 \sin^2 \theta = a^2(1 - \sin^2 \theta) = a^2 \cos^2 \theta$.
Thus

$$I = \int \frac{1}{a^2} \sec^2 \theta \, d\theta$$

$$= \frac{1}{a^2} \tan \theta + C \qquad \text{since } \frac{d}{d\theta} \tan \theta = \sec^2 \theta.$$

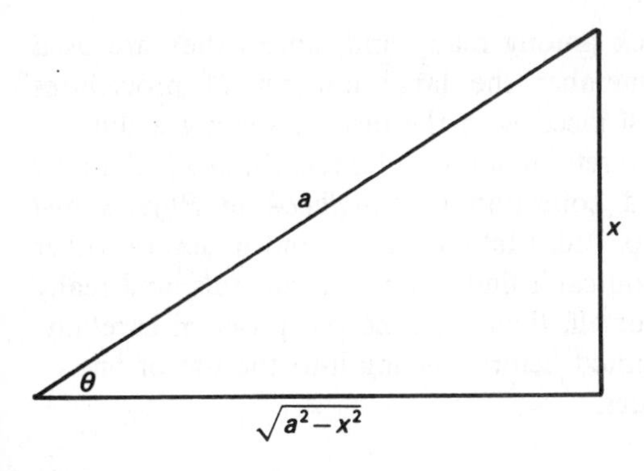

Figure 10.6

But if $\sin\theta = x/a$, then

$$\tan\theta = \frac{x}{\sqrt{a^2-x^2}}.$$

This is shown in Fig. 10.6. So, finally,

$$I = \frac{x}{a^2\sqrt{a^2-x^2}} + C.$$

10.11 SUMMARY OF INTEGRATION PROCEDURES

The techniques of integration listed here are only the ones most commonly encountered. Very many more are available. Consider, for example,

$$I = \int \frac{\mathrm{d}x}{x^2-x-6}.$$

Because

$$x^2-x-6=(x+2)(x-3)$$

the integral can be written as

$$I = \int \frac{1}{5}\left(\frac{1}{x-3} - \frac{1}{x+2}\right)\mathrm{d}x.$$

But

$$\int \frac{1}{x+2}\,\mathrm{d}x = \ln(x+2) + C.$$

Thus this device – breaking the fractional integrand into so-called partial fractions – allows the integral to be determined.

However, this is just one trick among many and, unless they are used regularly, it is difficult to remember the large number of procedures available. So if you find yourself faced with the task of solving a difficult integral my recommendation is to refer to a table of integrals, of which many are published. There is a good collection in *Handbook of Physics and Chemistry* (American Rubber Co. Ltd. 61st edn, 1989) and in several other laboratory reference books. If you can't find a table of integrals, and really are obliged to do the work yourself, then examine the problem carefully. Consider the strategy to be adopted before rushing into the use of one or other of the integration techniques.

EXAMPLES 10E

1. Integrate
   (a) $\dfrac{\sin x}{7+5\cos x}$;
   (b) $\dfrac{3x^2+2x}{(x^3+x^2+1)^3}$;
   (c) $x^2\cos(x^3)$.
2. Integrate
   (a) $(x+1)\ln(x+1)$;
   (b) $x^5(x^3+1)^5$ (Hint: $x^5=x^2\cdot x^3$).
3. Integrate
   (a) $\dfrac{1}{\sqrt{(x^2+4x+5)}}$; (Hint: use the form $(x+\alpha)^2+\beta^2$ then substitute $x+\alpha=\beta\tan\theta$, remembering the formula for $\tan^2\theta$.)
   (b) $\dfrac{x}{x^2+4x+5}$; (Hint: start as in (a), write the numerator as $x+\alpha-\alpha$ and split the integral in two.)
   (c) $\dfrac{x}{\sqrt{(x^2+4x+5)}}$.
4. Sketch the overlapping region of the curves $y^2=4x$ and $2x^2=y$ and calculate the area enclosed between them.
5. Evaluate $\int e^{3\theta}\cos 5\theta\,d\theta$.

Essentially I integrate the current export drive
And basically I'm viable from ten o'clock till five.

John Betjeman, *Executive*

11

Series, algebraic and trigonometric

A collection of numbers such as $4, x, A, 0, 19.7, i, n, \ldots,$ form a set. A collection arranged in a definite order, $a_1, a_2, a_3, \ldots,$ with a rule (overt or covert) for getting from one to the next form a sequence. If the successive terms in a sequence are added, the result is a series. This chapter is about a few of the series often encountered in elementary mathematics, with some attention also paid to the individual terms of the series, as if they were merely members of a sequence.

11.1 ARITHMETIC PROGRESSION (AP)

'What are the next numbers in (a) 7, 11, 15, 19,...
(b) 85, 76, 67, 58,...
(c) 1, 2, 5, 4, 9, 6, 13,...?'

(from an advertisement in a public place, supposedly offering an 'intelligence test'.) These sequences are examples of an arithmetic progression (AP). In (a) the numbers increase by 4 in going from one to the next; in (b) they decrease by 9; and in (c) two APs alternate, one increasing by 4 from one term to the next (viz. 1, 5, 9,...) and the other by 2 (viz. 2, 4, 6,...).

A sequence in AP with first term a and difference (often called the common difference) d from one term to the next is $a, a+d, a+2d, a+3d, \ldots$ Obviously, the nth term of the sequence is $a+(n-1)d$. The sum, S, of the first n terms of the sequence is given by

$$S=a+(a+d)+(a+2d)+\cdots+(a+(n-2)d)+(a+(n-1)d)$$

If this sequence is merely written in reverse order we have

$$S=(a+(n-1)d)+(a+(n-2)d)+\cdots+(a+2d)+(a+d)+a.$$

Adding these two series together, first term with first term, second with second, and so on up to nth term with nth, gives

$$2S=(2a+(n-1)d)+(2a+(n-1)d)+\cdots+(2a+(n-1)d).$$

That is, there are n terms, all of the same value, $(2a+(n-1)d)$. Hence, $2S=n(2a+(n-1)d)$ so the sum of the first n terms of a series with its terms in AP is

$$S=na+\frac{n(n-1)}{2}d. \tag{11.1}$$

Example 11(a) Find the sum of the first 50 odd numbers.

Answer: Use equation (11.1) with $a=1$, $d=2$, and $n=50$.
Then

$$S=50\cdot 1+\frac{50\cdot 49}{2}2=2500.$$

Example 11(b) In an AP, the ninth term is four times the second term and the sum of the first five terms is 50. What is the sequence?

Answer: Using the same symbolism as above, the ninth term is $a+8d$ and the second is $a+d$. Thus $a+8d=4(a+d)$. This gives

$$3a=4d \tag{a}$$

The sum to 5 terms is $5a+\frac{5\cdot 4}{2}d=50$, so

$$5a=50-10d \tag{b}$$

Substitute 5 times (a) in 2 times (b) giving $25a=100$. Therefore $a=4$, so $d=3$ and the sequence is 4, 7, 10, 13, ...

EXERCISES 11A

1. Find the tenth term and the sum of the first ten terms of the APs
   (a) $2+2\frac{1}{2}+3+\cdots$
   (b) $53+50+47+\cdots$
2. The first term of an AP is 2 and the sum of the first five terms is -40. What is the difference between successive terms?

11.2 GEOMETRIC PROGRESSION (GP)

In APs, the rule for getting from one term to the next is to add (or subtract) a constant quantity. In a geometric progression (GP) the rule is to multiply (or divide) by a constant quantity, the ratio (or common ratio) r, to get from one term to the next. Thus the terms of the sequence 2, 6, 18, 54, 162, ... are in GP, with ratio 3. Similarly, the terms 243, 81, 27, 9, ... are in GP with ratio $\frac{1}{3}$.

If a GP has a first term a and a ratio r, the sequence is $a, ar, ar^2, ar^3, \ldots,$ and the nth term is ar^{n-1}. The sum, S, of the first n terms is

$$S = a + ar + ar^2 + ar^3 + \cdots + ar^{n-2} + ar^{n-1}. \tag{11.2}$$

Multiplying S by r gives

$$rS = ar + ar^2 + ar^3 + \cdots + ar^{n-2} + ar^{n-1} + ar^n. \tag{11.3}$$

Subtracting equation (11.3) from equation (11.2) gives

$$S - rS = a - ar^n.$$

Hence the sum of the first n terms of a series with its terms in GP is

$$S = \frac{a(1-r^n)}{1-r} = \frac{a(r^n-1)}{r-1}. \tag{11.4}$$

(The two forms of equation (11.4) are identical, but it is more convenient to use the second if $r > 1$.)

The GP series $1+2+4+8+\cdots$, where the ratio of a term to its predecessor is 2, obviously grows without bound as the number of terms increases. But so also does the series $1+1.01+1.0201+1.030301+\cdots$, in which the ratio is 1.01. This conclusion follows from equation (11.4) because, if $r > 1$, then $r^n \to \infty$ as $n \to \infty$, even if $r-1$ is very small. If r is negative, terms of the sequence will be alternately positive and negative. If $r < -1$ then $|r^n| \to \infty$ as $n \to \infty$, i.e. the absolute value of r^n becomes immeasurably large as n increases without limit; but in this case, because successive values of n make the terms alternately positive and negative, the sum of the series oscillates between immeasurably large positive and negative values. Only if $|r| < 1$, i.e. only if $-1 < r < +1$, does the sum of a series of terms in GP tend to a definite finite value as the number of terms in the series is limitlessly increased, because in these circumstances $r^n \to 0$ as $n \to \infty$ and $S \to a/(1-r)$. A series which approaches a definite value like this, and also the sequence which consists just of the terms of the series, are said to **converge** or to show convergence.

In the case of $r = -1$, the value of r^n is alternately $+1$ and -1 for successive values of n. In this case the sum, S, does not approach a definite value, nor does it grow without limit; instead, it oscillates between the values a and 0. This is obvious since the series is reduced to $a - a + a - a + \cdots$. In the case of $r = +1$, equation (11.4) cannot be used, since division by zero is undefined. However, the series is reduced to $S = a + a + a + \cdots$ so the sum to n terms of this very uninteresting series is na and it tends to infinity with n.

Example 11(c) What is the tenth term and the sum of the first ten terms of the sequence 1, 2, 4, 8, ...?

Answer: The sequence is a GP with $a=1$ and $r=2$. The tenth term is $ar^9=512$. The sum of the first ten terms is

$$S=\frac{2^{10}-1}{2-1}=1023.$$

Example 11(d) Calculate the seventh term and the sum to seven terms and to infinity of
(a) $1+\frac{1}{2}+\frac{1}{4}+\frac{1}{8}+\cdots$
(b) $1-\frac{1}{2}+\frac{1}{4}-\frac{1}{8}+$

Answer:
(a) $a=1$ and $r=\frac{1}{2}$, so the seventh term is

$$1\cdot r^6=\frac{1}{64};$$

the sum to seven terms is

$$\frac{1-\frac{1}{128}}{1-\frac{1}{2}}=1.984;$$

and the sum to infinity is

$$\frac{1}{1-\frac{1}{2}}=2.$$

(b) $a=1$ and $r=-\frac{1}{2}$ so the seventh term is $-\frac{1}{64}$; the sum to seven terms is

$$\frac{1+\frac{1}{128}}{1+\frac{1}{2}}=0.672;$$

and the sum to infinity is

$$\frac{1}{1+\frac{1}{2}}=0.667.$$

Example 11(e) You take out a mortgage M for Y years at an extortionate annual rate of interest of E%. Derive a formula for your monthly payments.

Answer: Call the monthly payment P. The interest due on an amount A after one month is $EA/1200$ so the amount plus its interest is $A(1+E/1200)$. For convenience, write $k=(1+E/1200)$. Then one month after taking out the mortgage M, the amount owing is kM; but when the first payment is made $kM-P$ is owing. One month later, $k(kM-P)$ is owing, i.e. the amount owed after

one month is $kM-P$
two months is $k^2M-kP-P$
three months is $k^3M-k^2P-kP-P$
$\vdots \qquad \vdots$
n months is $k^nM-k^{n-1}P-k^{n-2}P-\cdots$.

Thus, after n months, the remaining debt D_n is given by

$$D_n=k^nM-P(1+k+k^2+k^3+\cdots+k^{n-1}).$$

The quantity in brackets is a series with its terms in GP; there are n terms and the sum of them is $(k^n-1)/(k-1)$. Hence

$$D_n=k^nM-P\frac{k^n-1}{k-1}.$$

But, by the terms of the mortgage, $D_n=0$ when $n=12Y$. Thus the required formula is

$$P=\frac{k^{12Y}(k-1)}{k^{12Y}-1}M \qquad \text{where} \qquad k=1+\frac{E}{1200}.$$

(NB: many companies which offer mortgages charge interest for the whole year on the debt outstanding at the beginning of the year and give no credit for the the reduction of the debt by the monthly payments made during the year. If you are repaying a mortgage, use the formula to check whether the payments you are making are correct.)

EXERCISES 11B

1. Find the sixth term, the sum to six terms and, if it is finite, the sum to infinity of the series formed of terms in the sequences
   (a) 1.000, 0.900, 0.810, 0.729, ...
   (b) 1.000, -1.100, 1.210, -1.331, ...
   (c) 81, 27, 9, 3, ...
2. Sum to infinity the series $1+2x+4x^2+8x^3+\cdots$. For what range of values of x is your answer valid?

11.3 SUMMATION OF SERIES BY THE METHOD OF DIFFERENCES

The method of differences can be used to sum several series which themselves do not have much in common. It involves rewriting each term of the series as one quantity minus another quantity in such a way that, when the terms are added, most of the intermediate quantities cancel each other out. The method can best be demonstrated by way of examples.

Example 11(f) Find the sum of the first n terms of the series $1\cdot3+3\cdot5+5\cdot7+\cdots$ where each term is the product of two numbers, as shown.

Answer: The first number of the pair in each term form the sequence 1, 3, 5,... so, calling the nth term u_n, we can write $u_n=(2n-1)(2n+1)$. Now consider a related series formed of terms v_n, where

$$v_n=(2n-1)(2n+1)(2n+3).$$

It follows (putting $n-1$ in place of n) that

$$v_{n-1}=(2n-3)(2n-1)(2n+1).$$

Then

$$\begin{aligned} v_n-v_{n-1} &= (2n-1)(2n+1)((2n+3)-(2n-3)) \\ &= 6(2n-1)(2n+1) \\ &= 6u_n. \end{aligned}$$

so

$$u_n=\tfrac{1}{6}(v_n-v_{n-1}).$$

Similarly,

$$\begin{aligned} u_{n-1} &= \tfrac{1}{6}(v_{n-1}-v_{n-2}), \\ u_{n-2} &= \tfrac{1}{6}(v_{n-2}-v_{n-3}), \\ &\vdots \\ u_2 &= \tfrac{1}{6}(v_2-v_1), \\ u_1 &= \tfrac{1}{6}(v_1-v_0). \end{aligned}$$

Adding up the two sides of this ladder of equations, on the l.h.s. there results $S=u_1+u_2+u_3+\cdots+u_n$; and on the r.h.s., going up the ladder, one term just cancels out another in the equation above, all except for one term on the bottom rung of the ladder and one on the top. What is left is

$$\sum_i u_i=S=\tfrac{1}{6}(v_n-v_0).$$

Putting $n=0$ in the equation for v_n gives $v_0=-3$. Hence

$$S=\tfrac{1}{6}((2n-1)(2n+1)(2n+3)+3).$$

Example 11(g) Find the sum of the first n integers.

Answer: The problem is to find $S=1+2+3\cdots+n$. Here put $u_n=n$ so that

$S=\sum_i u_i$ and $v_n=n(n+1)$ so that $v_{n-1}=(n-1)n$. Thus,

$$v_n-v_{n-1}=n((n+1)-(n-1))=2u_n.$$

Forming a ladder as in example 11(f) and adding each side leads to

$$\begin{aligned}S&=\tfrac{1}{2}(v_n-v_0)\\&=\tfrac{1}{2}(n(n+1)-0).\end{aligned}$$

Thus the required sum is $\frac{1}{2}n(n+1)$. (This result could also have been obtained as the sum of an AP with $a=1$ and $d=1$ in equation (11.1).

Example 11(h) Sum to n terms the series

$$S=\frac{1}{1\cdot 2\cdot 3}+\frac{1}{2\cdot 3\cdot 4}+\frac{1}{3\cdot 4\cdot 5}+\cdots.$$

Answer: In this case,

$$u_n=\frac{1}{n(n+1)(n+2)}.$$

Consider

$$v_n=\frac{1}{n(n+1)} \qquad \text{and} \qquad v_{n+1}=\frac{1}{(n+1)(n+2)}.$$

Then

$$v_n-v_{n+1}=\frac{(n+2)-\mathrm{n}}{n(n+1)(n+2)}=2u_n.$$

Forming a ladder as in example 11(f) and adding each side leads to

$$\begin{aligned}S=\sum_i u_i&=\tfrac{1}{2}(v_1-v_{n+1})\\&=\frac{1}{2}\left(\frac{1}{1\cdot 2}-\frac{1}{(n+1)(n+2)}\right).\end{aligned}$$

Hence

$$S=\frac{n(n+3)}{4(n+1)(n+2)}.$$

EXERCISE 11C

Use the methods of examples 11(f)–11(h) to sum to ten terms the series

(a) $1\cdot 2+2\cdot 3+3\cdot 4+\cdots$

(b) $\frac{1}{1\cdot 2}+\frac{1}{2\cdot 3}+\frac{1}{3\cdot 4}+\cdots$

11.4 MACLAURIN'S SERIES

Let y be a function of x, say $y=f(x)$, where the function might be algebraic or trigonometric or a mixture of the two. A theorem by Maclaurin states that y can be expressed as a power series (a polynomial of unlimited degree) in x, that is, we can put

$$f(x)=a_0+a_1x+a_2x^2+a_3x^3+\cdots+a_nx^n+\cdots, \tag{11.5}$$

and explains how to solve the problem of calculating the coefficients a_0, a_1, $a_2, \ldots$

The calculation of the first coefficient is particularly easy. Put $x=0$ and all the terms on the r.h.s. of equation (11.5) vanish except the first, leaving

$$a_0=f(0).$$

Now differentiate equation (11.5) with respect to x, giving

$$f'(x)=a_1+2a_2x+3a_3x^2+4a_4x^3+\cdots$$

Again put $x=0$ and again on the r.h.s. all terms except the first vanish, leaving

$$a_1=f'(0).$$

Differentiate equation (11.5) a second time with respect to x, giving

$$f''(x)=2a_2+3\cdot 2a_3x+4\cdot 3a_4x^2+\cdots+n\cdot(n-1)a_nx^{n-2}+\cdots$$

This time, when $x=0$, what is left gives

$$a_2=\frac{1}{2}f''(0)=\frac{1}{2!}f''(0).$$

Repeat the whole procedure once more to get

$$a_3=\frac{1}{3!}f'''(0),$$

and again to get

$$a_4=\frac{1}{4!}f''''(0),$$

and repeat in this way to accumulate the evaluated coefficients in the series:

$$f(x)=f(0)+f'(0)\cdot x+\frac{1}{2!}f''(0)x^2+\frac{1}{3!}f'''(0)x^3+\frac{1}{4!}f''''(0)x^4+\cdots \tag{11.6}$$

Equation (11.6) is Maclaurin's series. It can be used to find a wide variety of series – for example, the binomial series (cf. Chapter 3) when $f(x)=$

$(1+x)^p$ – but we limit its application here just to two functions, $\sin x$ and $\cos x$.

11.5 SERIES FOR THE TRIGONOMETRIC FUNCTIONS

To expand the function $f(x)=\sin x$, note that $f'(x)=\cos x$, $f''(x)=-\sin x$, $f'''(x)=-\cos x$, $f''''(x)=\sin x$, etc. But, as we know, $\sin 0=0$ and $\cos 0=1$. Substituting these values in equation (11.6) gives the result that

$$\sin x = x - \frac{x^3}{3!} + \frac{x^5}{5!} - \frac{x^7}{7!} + \cdots \tag{11.7}$$

That is, the sine function can be expressed as a power series, but containing only the odd powers, each divided by its corresponding factorial, and with terms alternately positive and negative.

Similarly, if $f(x)=\cos x$, then $f'(x)=-\sin x$, $f''(x)=-\cos x$, $f'''(x)=\sin x$, $f''''(x)=\cos x$, etc., and in this case,

$$\cos x = 1 - \frac{x^2}{2!} + \frac{x^4}{4!} - \frac{x^6}{6!} + \cdots \tag{11.8}$$

Thus the cosine series is very like the sine series except that it contains only even powers.

The series for sine and cosine confirm or reveal a number of properties of the functions, several of which have been met before in Chapter 4. For example, since $(-a)^n$ is positive if n is an even integer and negative if it is odd, equations (11.7) and (11.8) show that

$$\sin(-x) = -\sin x$$
$$\cos(-x) = \cos x.$$

If $f(-x)=-f(x)$, the function is said to be odd, so $\sin x$ is an odd function; and if $f(-x)=f(x)$, the function is said to be even, so $\cos x$ is an even function. From equations (11.7) and (11.8) it also follows that as x becomes very small $\sin x \to x$ and $\cos x \to 1$; these were results intuitively inferred before in Section 4.6 but equations (11.7) and (11.8) allow the accuracy of the approximation to be checked.

Compare the series in equations (11.7) and (11.8) with the series for the exponential function of equation (3.7), i.e.

$$e^x = 1 + x + \frac{x^2}{2!} + \frac{x^3}{3!} + \frac{x^4}{4!} + \frac{x^5}{5!} + \cdots.$$

There is clearly a close correspondence, but it is of the 'nearly but not quite' kind. Now look at the series for e^{ix} where i is the imaginary number and ix is written in place of x in the exponential series. Remember that $i^2=-1$, so that $i^3 = -i$, and $i^4 = +1$, $i^5 = 1$, $i^6 = -1$, $i^7 = -i$, etc. Then

$$e^{ix} = 1 + ix - \frac{x^2}{2!} - i\frac{x^3}{3!} + \frac{x^4}{4!} + i\frac{x^5}{5!} - \cdots.$$

Collect all the real terms together and all the imaginary terms together. This gives

$$e^{ix} = \left(1 - \frac{x^2}{2!} + \frac{x^4}{4!} - \frac{x^6}{6!} + \cdots\right) + i\left(x - \frac{x^3}{3!} + \frac{x^5}{5!} - \frac{x^7}{7!} + \cdots\right)$$

Compare this with equations (11.7) and (11.8) and see that a result of great subsequent importance has been established:

$$e^{ix} = \cos x + i \cdot \sin x. \tag{11.9}$$

Using equation (11.9) we can confirm de Moivre's theorem of section 5.6 (equation 5.4) without recourse to the Argand diagram. For

$$(\cos\theta + i\sin\theta)^n = (e^{i\theta})^n = e^{in\theta}$$
$$= \cos n\theta + i\sin n\theta.$$

From equation (11.9) it also follows that $e^{i\pi} = -1$, a constraint which perhaps helps to contain the elusiveness of the three numbers e, i, and π.

If equation (11.9) is inverted it gives

$$e^{-ix} = \cos x - i\sin x. \tag{11.10}$$

Adding equations (11.9) and (11.10) and making a small rearrangement gives

$$\cos x = \tfrac{1}{2}(e^{ix} + e^{-ix}); \tag{11.11}$$

and subtracting these equations gives similarly

$$\sin x = \frac{1}{2i}(e^{ix} - e^{-ix}). \tag{11.12}$$

There is now a set of results, in equations (11.9)–(11.12), which allows conversion both ways between the trigonometric functions and the exponential function of imaginary arguments.

Example 11(i) Express $\cos^3 x$ as a power series in x.

Answer:

$$\cos^3 x = \tfrac{1}{8}(e^{ix} + e^{-ix})^3$$
$$= \tfrac{1}{8}(e^{3ix} + 3e^{ix} + 3e^{-ix} + e^{-3ix})$$
$$= \tfrac{1}{4}\cos 3x + \tfrac{3}{4}\cos x.$$

This result can also be reached by writing:

$$\begin{aligned}\cos^3 x &= (\cos x)(\cos^2 x)\\ &= (\cos \mathrm{x})(\cos 2x + 1)\\ &= \tfrac{1}{2}[\tfrac{1}{2}(\cos 3x + \cos x) + \cos x])\end{aligned}$$

$\therefore$

$$\cos^3 x = 1 - \left(\frac{3^2}{4} + \frac{3}{4}\right)\frac{x^2}{2!} + \left(\frac{3^4}{4} + \frac{3}{4}\right)\frac{x^4}{4!} - \left(\frac{3^6}{4} + \frac{3}{4}\right)\frac{x^6}{6!} + \left(\frac{3^8}{4} + \frac{3}{4}\right)\frac{x^8}{8!} + \cdots$$

$$= 1 - \frac{3}{2!}x^2 + \frac{21}{4!}x^4 - \frac{183}{6!}x^6 + \frac{1641}{8!}x^8 + \cdots$$

EXERCISES 11D

1. Express $\sin^2 2x$ as a power series in x.
2. Show that the first three terms of the sine and cosine series are sufficient to calculate $\sin \pi/6$ and $\cos \pi/6$ to an accuracy of about 1 part in 10^4.

11.6 THE HYPERBOLIC FUNCTIONS

Equations (11.11) and (11.12) express the trigonometric functions in terms of the exponential function of imaginary arguments. What would be the situation if the arguments of the exponential function were real? For this case the hyperbolic functions have been defined. Here are the simplest and most frequently encountered of them.

Hyperbolic cosine of x, written as $\cosh x$ (and read as 'cosh x'):

$$\begin{aligned}\cosh x &= \tfrac{1}{2}(e^x + e^{-x})\\ &= 1 + \frac{x^2}{2!} + \frac{x^4}{4!} + \frac{x^6}{6!} + \cdots \end{aligned} \tag{11.13}$$

Hyperbolic sine of x, written as $\sinh x$ (and read as 'shine x'):

$$\begin{aligned}\sinh x &= \tfrac{1}{2}(e^x - e^{-x})\\ &= x + \frac{x^3}{3!} + \frac{x^5}{5!} + \frac{x^7}{7!} + \cdots \end{aligned} \tag{11.14}$$

Hyperbolic tangent of x, written as $\tanh x$ (and read as 'tanj x'):

$$\tanh x = \frac{\sinh x}{\cosh x} = \frac{e^x - e^{-x}}{e^x + e^{-x}}.$$

Without going into details, it suffices to say that other hyperbolic functions are defined by analogy with the trigonometric functions, e.g. $\operatorname{sech} x = 1/\cosh x$, etc.

Example 11(j) Show that

(a) $\cosh^2\theta - \sinh^2\theta = 1$;
(b) $\sinh 2\theta = 2\sinh\theta\cosh\theta$;
(c) $\cosh(A+B) = \cosh A\cosh B + \sinh A\sinh B$.

Answer: Make direct use of equations (11.13) and (11.14)

(a)
$$\begin{aligned}\cosh^2\theta - \sinh^2\theta &= \tfrac{1}{4}(e^{\theta}+e^{-\theta})^2 - \tfrac{1}{4}(e^{\theta}-e^{\theta})^2\\ &= \tfrac{1}{4}[(e^{2\theta}+2+e^{-2\theta})-(e^{2\theta}-2+e^{-2\theta})]\\ &= \tfrac{1}{4}\cdot 4 = 1.\end{aligned}$$

(b)
$$\begin{aligned}2\sinh\theta\cosh\theta &= 2\cdot\tfrac{1}{2}(e^{\theta}-e^{-\theta})\tfrac{1}{2}(e^{\theta}+e^{-\theta})\\ &= \tfrac{1}{2}(e^{2\theta}-e^{-2\theta})\\ &= \sinh 2\theta.\end{aligned}$$

(c)
$$\begin{aligned}\cosh A\cosh B &= \tfrac{1}{2}(e^{A}+e^{-A})\tfrac{1}{2}(e^{B}+e^{-B})\\ &= \tfrac{1}{4}(e^{A+B}+e^{-A-B}+e^{A-B}+e^{-A+B})\\ \sinh A\sinh B &= \tfrac{1}{2}(e^{A}-e^{-A})\tfrac{1}{2}(e^{B}-e^{-B})\\ &= \tfrac{1}{4}(e^{A+B}+e^{-A-B}-e^{A-B}-e^{-A+B})\end{aligned}$$

$$\therefore \qquad \begin{aligned}\cosh A\cosh B + \sinh A\sinh B &= \tfrac{1}{4}(2e^{A+B}+2e^{-A-B})\\ &= \cosh(A+B).\end{aligned}$$

11.7 GENERAL NOTES ON THE HYPERBOLIC FUNCTIONS

These functions clearly have a close similarity to the trigonometric functions. This is seen in the parallel definitions revealed by comparing equations (11.13) and (11.14) with equations (11.11) and (11.12) and by comparing the identities of example 11(j) and exercises 11E with the corresponding identities for the trigonometric functions in Table 4.3. However, there are also some crucial differences between the two sets of functions.

1. All of the trigonometric functions are periodic; none of the hyperbolic functions are.
2. The sine and cosine functions lie between $+1$ and -1 and the tangent function is unlimited; whereas the hyperbolic sine and the hyperbolic cosine are unlimited (increasing as $\pm e^{\theta}$ as $|\theta|$ becomes very large) whereas the hyperbolic tangent is constrained to lie between $+1$ and -1.

3. In identities such as those of example 11(j), the sign is sometimes the same as in the corresponding trigonometric identities, but is sometimes different. (Invariably, it is the sign in front of $\sinh^2\theta$ which reverses, and this is known as Osborn's rule.)

EXERCISES 11E

1. Show that
   (a) $\cosh^2 A + \sinh^2 A = \cosh 2A$;
   (b) $\tanh^2 A = 1 - \operatorname{sech}^2 A$.
2. Show that
   (a) $\dfrac{\mathrm{d}}{\mathrm{d}x}(\sinh x) = \cosh x$;

   (b) $\dfrac{\mathrm{d}}{\mathrm{d}x}(\cosh x) = \sinh x$;

   (c) $\dfrac{\mathrm{d}}{\mathrm{d}x}(\tanh x) = \operatorname{sech}^2 x$.

12

Differential equations

Equations involving variables (which, as usual, we can call x and y), including one or more derivatives and perhaps also terms containing x and y themselves, are called differential equations. Thus the many equations in Chapter 10, of the form $\mathrm{d}y/\mathrm{d}x = h(x)$, where y appears only on the l.h.s. within the derivative, are differential equations. Extracting y as a function of x from one of these equations was regarded there as just a problem of integration; but it could equally well have been regarded as the problem of solving the equation, the solution in these cases being the function of x which determines y and satisfies the equation. However, with most differential equations, various procedures of greater or lesser complexity must first be completed before a final integration can be made to reveal the direct relation between the variables.

Differential equations and their solutions lie at the core of quantitative science. This is because a scientific hypothesis being set up to explain a dynamic situation is typically couched in terms of the dependence of rates of change on the various factors influencing the situation, and when this is expressed symbolically a differential equation results. Rates of change are derivatives, so a statement of how a rate of change is affected produces an equation of the kind we shall examine here.

The **order** of a differential equation is that of the highest derivative in it. The **degree** of such an equation is the power to which the highest derivative is raised. Thus an equation like

$$\left(\frac{\mathrm{d}^3 y}{\mathrm{d}x^3}\right)^2 + (2 \sin x^5 + 1)\left(\frac{\mathrm{d}^2 y}{\mathrm{d}x^2}\right)^3 + 3\frac{\mathrm{d}y}{\mathrm{d}x} + 4y^4 = 0$$

is a third-order differential equation (since it involves $(\mathrm{d}^3 y/\mathrm{d}x^3)$) of the second degree (since $\mathrm{d}^3 y/\mathrm{d}x^3$ is squared). However, in the limited look at differential equations to be given here, we shall be concerned exclusively with first-degree (or linear) equations and only with equations of first or second order. Furthermore, this survey is limited to a few of the most frequently encountered types; in each case, note the pattern of the type of equation so that if you meet it again you will know what method of solution to employ.

12.1 EQUATIONS OF THE FIRST ORDER AND FIRST DEGREE: VARIABLES SEPARABLE

If the equation to be solved is of the form

$$h(y)\frac{\mathrm{d}y}{\mathrm{d}x}=k(x) \tag{12.1}$$

then it can be immediately integrated. For suppose that (differentiating with respect to x)

$$\frac{\mathrm{d}}{\mathrm{d}x}K(x)=k(x)$$

and (differentiating with respect to y)

$$\frac{\mathrm{d}}{\mathrm{d}y}H(y)=h(y);$$

then (differentiating with respect to x again)

$$\frac{\mathrm{d}}{\mathrm{d}x}H(y)=h(y)\frac{\mathrm{d}y}{\mathrm{d}x}.$$

Now integrate both sides of equation (12.1) with respect to x. But

$$\int h(y)\frac{\mathrm{d}y}{\mathrm{d}x}\,\mathrm{d}x=\int h(y)\,\mathrm{d}y$$

using integration by substitution as in section 10.10. Hence equation (12.1) becomes

$$\int h(y)\,\mathrm{d}y=\int k(x)\,\mathrm{d}x. \tag{12.2}$$

The procedure that has led to this point is called the method of separation of variables because equation (12.1) has been rearranged so that in equation (12.2) all the quantities containing y are on one side of the equals sign and all the quantities containing x are on the other. Equation (12.2) looks as if it came from equation (12.1) after multiplying through by $\mathrm{d}x$ and putting the $\int$ sign in front of each side. This is a practical way of looking at the method, but mathematically dubious (cf. Section 10.2). Nevertheless, mathematicians allow one to go straight from equation (12.1) to equation (12.2) provided that abracadabra words like 'by the method of separation of variables' are muttered.

From equation (12.2) it then follows that

$$H(y)=K(x)+C.$$

Actually, both integrations of equation (12.2) provide an arbitrary constant, but precisely because the constants are arbitrary, they can be collected on to one side as just one such constant.

Example 12(a) Solve the equation $\dfrac{dy}{dx}=x^3/y^2$.

Answer: Rearrange the equation as $y^2\,dy/dx=x^3$ and integrate:

$$\int y^2\,dy=\int x^3\,dx$$

$$\therefore \qquad \tfrac{1}{3}y^3=\tfrac{1}{4}x^4+C'$$

or

$$4y^3=3x^4+C.$$

(NB: as stated above, since the constant of integration is arbitrary, it need be placed only on one side, and it is simpler, but no less general, to write C rather than $12C'$.)

Example 12(b) If a mixture of two liquids, A and B, is heated it is found that the relative rate at which one evaporates relative to the other, i.e. dA/dB, is proportional to the ratio of the volumes of liquids present. Find the law relating the volumes of the two liquids.

Answer: From the observations,

$$\frac{dA}{dB}=k\frac{A}{B},$$

where k is the constant of proportionality. Hence,

$$\int \frac{1}{A}\,dA=\int k\frac{1}{B}\,dB$$

$$\therefore \qquad \ln A=k\ln B+C$$

$$=\ln B^k+C.$$

Thus,

$$\ln\left(\frac{A}{B^k}\right)=C$$

or

$$\frac{A}{B^k} = e^C = K \text{ (say).}$$

So $A = KB^k$, where K is now the arbitrary constant.

EXERCISE 12A

Solve

(a) $xy\frac{dy}{dx} = 1 + x$;

(b) $\sin 2y\frac{dy}{dx} = [\cos(x+y) + \cos(x-y)] \sin 2x$.

12.2 EQUATIONS OF THE FIRST ORDER AND FIRST DEGREE (LINEAR): INTEGRATING FACTOR

A type of equation which very frequently appears in all branches of science is of the form

$$\frac{dy}{dx} + r(x) \cdot y = s(x). \tag{12.3}$$

That is, there are three terms of which one is just the first derivative of y with respect to x, a second is y multiplied by a function r of x, and the third is a function s of x alone. Incidentally, either r or s or both can be constants. There is no point in going through the analysis of the method of solving this equation. Here are the rules for arriving at the answer.

1. Form $R(x) = \int r(x)\, dx$, with zero integrating constant. (NB: the inverse of this procedure gives $dR/dx = r(x)$.)
2. Write R for $R(x)$ and multiply through equation (12.3) by e^R, called the integrating factor.
3. Integrate both sides of the equation with respect to x.
4. Tidy up the result, which usually means dividing through by the integrating factor.

This technique can perhaps be best understood by first looking at it in reverse. Suppose that the end result is

$$y = S(x) + Ce^{-R}$$

where C is the arbitrary constant of the integration. Multiply through by e^R (the reverse of step 4.) to get

$$e^R y = e^R S(x) + C.$$

Now differentiate with respect to x (the reverse of step 3.) to get

$$e^R \frac{dy}{dx} + e^R \frac{dR}{dx} y = \frac{d}{dx}(e^R S(x))$$

or

$$e^R \frac{dy}{dx} + e^R r(x) y = e^R s(x)$$

since $dR/dx = r(x)$ and writing

$$s(x) = S(x) + \frac{d}{dx} S(x).$$

Dividing through by e^R, the integrating factor, and so reversing step 2., now returns us to the starting point, equation (12.3).

The integrating factor is thus a device which you must introduce into the equation to permit its solution. Its presence allows the l.h.s. of equation (12.3) to be integrated in a way which separates the variables.

Example 12(c) Solve $dy/dx + 2xy = 2x$.

Answer: Following the procedure laid down,

1. $\int 2x\,dx = x^2$, so multiply through by e^{x^2} to get:
2. $e^{x^2} \dfrac{dy}{dx} + 2x\,e^{x^2} y = 2xe^{x^2}$, of which the l.h.s. can be written as $\dfrac{d}{dx}(y\,e^{x^2})$.
3. Integrating,

$$\int \frac{d}{dx}(ye^{x^2})\,dx = \int 2xe^{x}\,dx.$$

Hence $ye^{x^2} = e^{x^2} + C$ so that $y = 1 + Ce^{-x^2}$.

Example 12(d) Solve $x(x+1)\,dy/dx + y = (x+1)^3$.

Answer: Divide through by $x(x+1)$ giving

$$\frac{dy}{dx} + \frac{y}{x(x+1)} = \frac{(x+1)^2}{x}.$$

Now $$\int \frac{dx}{x(x+1)} = \int \left[\frac{1}{x} - \frac{1}{x+1}\right] dx = \ln x - \ln(x+1) = \ln \frac{x}{x+1}.$$

But $e^{\ln p} = p$, so the integrating factor is $x/(x+1)$. Multiplying through by the integrating factor gives

$$\frac{x}{x+1}\frac{dy}{dx} + \frac{y}{(x+1)^2} = (x+1).$$

Integrating,
$$\frac{xy}{x+1}=\int (x+1)\,dx$$
$$=\tfrac{1}{2}(x+1)^2+C'.$$

Hence,
$$y=\frac{(x+1)^3+C(x+1)}{2x}.$$

Equations solved by use of this technique of introducing an integrating factor appear so commonly that it is worth looking at some simple examples of the kind of situations in which they are found. They characteristically follow from statements in which rates of change are said to depend on experimental or observed conditions of one kind or another. Note the contrast between the typically long account of the problem expressed in words and its pithy appearance as a differential equation (recall the beginning of Chapter 2).

Example 12(e) An oil bath is heated at a constant rate, R, but loses heat in proportion to the difference between its temperature, T, and the ambient temperature, T_0 (Newton's law of cooling). If the bath is initially at temperature T_0 when the heater is turned on, find how its temperature increases with time.

Answer: The rate of change of temperature with time, t, is $\mathrm{d}T/\mathrm{d}t$. The rate of rise of temperature is R. The rate of fall of temperature is $F(T-T_0)$, where F is the constant of proportionality.
(NB: we can conclude immediately that the temperature will rise to an equilibrium value T_e at which the rate of fall equals the rate of rise, i.e. $F(T_e-T_0)=R$ so that $T_e=T_0+R/F$, but this gives no indication of the time required to reach T_e.)

The differential equation governing the situation is the symbolic form of the statement that the rate of change of temperature equals the rate of rise less the rate of fall:

$$\frac{\mathrm{d}T}{\mathrm{d}t}=R-F(T-T_0)$$

$$\therefore \qquad \frac{\mathrm{d}T}{\mathrm{d}t}+FT=R+FT_0.$$

The integrating factor is $\mathrm{e}^{\int F\mathrm{d}t}=\mathrm{e}^{Ft}$. Hence,

$$T\mathrm{e}^{Ft}=\int (R+FT_0)\,\mathrm{e}^{Ft}\,\mathrm{d}t$$

$$=\frac{R+FT_0}{F}\mathrm{e}^{Ft}+C.$$

To evaluate the arbitrary constant, C, it is necessary to use the information given, namely that at $t=0$ (i.e. at the start of heating) $T=T_0$. Hence

$$T_0=\frac{R+FT_0}{F}+C \qquad \text{since } e^0=1.$$

So $C=-R/F$. Thus

$$T=\frac{R+FT_0}{F}-\frac{R}{F}e^{-Ft}$$

$$=T_0+\frac{R}{F}(1-e^{-Ft}).$$

This equation shows that $T=T_0$ at $t=0$ as required and that, at very large t, when e^{-Ft} is negligibly small, T tends to T_0+R/F, which we previously inferred. The change of temperature with time is shown in Fig. 12.1.

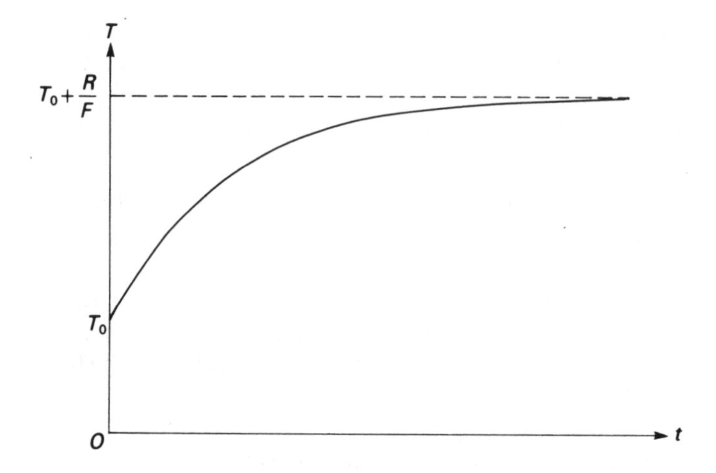

Figure 12.1

Example 12(f) A patient is treated by the venous injection of a quantity of drug in an inactive form. Metabolic reactions in the blood convert the drug at a constant rate k_1 into an active form which then is removed by the liver at a constant rate k_2. Supposing that negligible time is required for the initial injection of the drug to achieve a uniform concentration D_0 in the blood, find the time course in the blood of the concentration of the active form.

Answer: Let D be the concentration of the inactive form of the drug and A be the concentration of the active form. Then the rate of change of D is given by

$$\frac{dD}{dt} = -k_1 D.$$

This equation can be solved by the procedure of separable variables, i.e.

$$\int \frac{dD}{D} = \int -k_1 \, dt$$

giving

$$\ln D = -k_1 t + C,$$

so that

$$D = e^{-k_1 t + C}.$$

But $D = D_0$ when $t = 0$, so $e^C = D_0$ and $D = D_0 e^{-k_1 t}$.
The rate of change of the concentration of the active form of the drug equals its rate of formation less its rate of removal.

The rate of formation of the active form of the drug equals the rate at which the inactive form is converted; so rate of formation is

$$-\frac{dD}{dt} = k_1 D_0 \, e^{-k_1 t}.$$

The rate of removal is $-k_2 A$. Hence,

$$\frac{dA}{dt} = k_1 D_0 \, e^{-k_1 t} - k_2 A,$$

i.e.

$$\frac{dA}{dt} + k_2 A = k_1 D_0 \, e^{-k_1 t}.$$

The integrating factor is $e^{k_2 t}$. Hence,

$$A \, e^{k_2 t} = \int k_1 D_0 \, e^{(k_2 - k_1)t} \, dt \tag{12.4}$$

$$= \frac{k_1 D_0}{k_2 - k_1} e^{(k_2 - k_1)t} + C.$$

But $A = 0$ at $t = 0$ so $C = -k_1 D_0/(k_2 - k_1)$. Thus

$$A = \frac{k_1 D_0}{k_2 - k_1} (e^{-k_1 t} - e^{-k_2 t}). \tag{12.5}$$

(This assumes that $k_2 > k_1$. If $k_2 < k_1$, the terms in the bracket and the terms in the denominator of equation (12.5) can be interchanged. In the unlikely

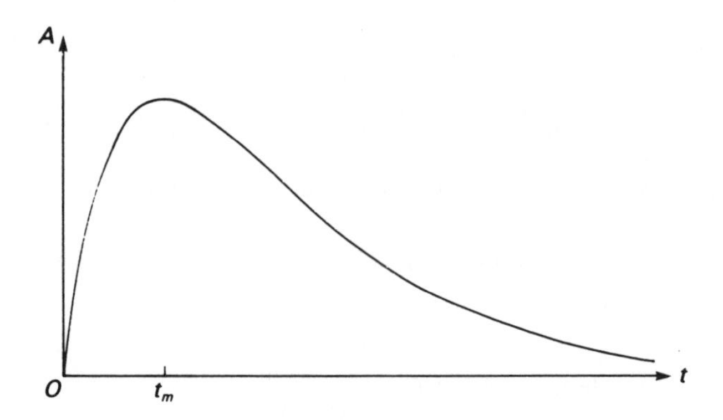

Figure 12.2

event of $k_1 = k_2$, the exponential term on the r.h.s. of equation (12.4) vanishes and the integral yields $k_1 D_0 t + C$ leading to $A = k_1 D_0 t\,e^{-k_1 t}$). The change of A with time is shown in Fig. 12.2. The concentration A is a maximum when $dA/dt = 0$, i.e. when

$$k_1 D_0 (k_2 e^{-k_2 t} - k_1 e^{-k_1 t}) = 0.$$

This occurs at a time, t_m, when the bracket is zero, i.e. when

$$k_2 e^{-k_2 t_m} = k_1 e^{-k_1 t_m}$$

so

$$\ln k_2 - k_2 t_m = \ln k_1 - k_1 t_m.$$

therefore

$$t_m = \frac{\ln(k_2/k_1)}{k_2 - k_1}.$$

(In the case where $k_2 = k_1$, the curve of A versus t is similar to that of Fig. 12.2 and the maximum occurs at $t_m = 1/k_1$.) Pharmaceutical manufacturers take great care in controlling the formulations of their products so that k_1 and k_2 have values such as to make t_m and the width of the peak in Fig. 12.2 as appropriate as possible to the condition being treated.

EXERCISES 12B

1. Solve

(a) $x^2 \dfrac{dy}{dx} - y = x^3 e^{(x^2-1)/x}$;

(b) $\dfrac{dx}{d\theta} + x \cot\theta = \operatorname{cosec}\theta \cot\theta$;

(c) $\dfrac{dx}{d\theta}+x\sin 2\theta=\theta\, e^{\cos^2\theta}$.

2. Within the city of Hamelin, after the rats left, the mice began to breed at a rate which would have increased their population, m, by a fraction B per annum. However, the mayor and corporation decreed that each household must provide one cat and one mousetrap, so that the citizens then trapped T mice each year and each cat caught, on average, a number of mice such that the mouse population was stabilized at M_0. After a new mayor was elected on an anti-rodent ticket, each household was required to provide a second cat. There was no change in the breeding rate of the mice, the rate at which the mice were trapped or the average number of mice caught by each cat. If there were H households,
   (a) what was the average number of mice caught by each cat in a year? (i.e. what number of caught mice per cat made $dm/dt=0$ under the old mayor?)
   (b) what happened to the mouse population, m, when the cat population doubled?

12.3 EQUATIONS OF THE SECOND ORDER AND FIRST DEGREE (LINEAR): CONSTANT COEFFICIENTS

The equations to be briefly examined here are of the kind

$$a\frac{d^2y}{dx^2}+b\frac{dy}{dx}+cy=f(x) \tag{12.6}$$

where a, b, and c are constants and are real. Equations of this form have been of immense importance in physics. In biology they occur less frequently than the first-order linear equations of the kind dealt with in the previous section, but are by no means rare.

Whereas the solutions of first-order linear equations like equation (12.3) contain just one arbitrary constant, the solutions of second-order linear equations contain two. As in the case of first-order equations, information about initial conditions, or about some other appropriate point, are required if values are to be given to these constants. And apart from involving two arbitrary constants, an equation like equation (12.6) has a solution which is composed of two distinct parts. This can be seen if it is supposed that u is a function of x which satisfies equation (12.6)

$$a\frac{d^2u}{dx^2}+b\frac{du}{dx}+cu=f(x). \tag{12.7}$$

But now look at $y=u+v$, where v is also a function of x. Substituting for y

in equation (12.6), using

$$\frac{\mathrm{d}}{\mathrm{d}x}(u+v)=\frac{\mathrm{d}u}{\mathrm{d}x}+\frac{\mathrm{d}v}{\mathrm{d}x},$$

etc. and subtracting equation (12.7) from the result, what is left states that

$$a\frac{\mathrm{d}^2v}{\mathrm{d}x^2}+b\frac{\mathrm{d}v}{\mathrm{d}x}+cv=0. \tag{12.8}$$

If v is a solution of equation (12.8) then:

1. $y=u+v$ is the **complete solution** of equation (12.6);
2. u is called the **particular integral** of equation (12.6);
3. v is called the **complementary function** of u.

The arbitrary constants associated with the solution of equation (12.6) are found only in the complementary function, v; there are no arbitrary constants in the particular integral, u.

In solving equation (12.6), obtaining the complementary function is easy – there are rules to be mechanically applied and they produce the answer without fail. On the other hand, obtaining the particular integral can be very difficult, but only easy cases are dealt with here. For both parts, the account which follows states what to do to get the answer. There is just a hint of where the method comes from, either enough of a hint to quell your suspicions or enough to show you what to look out for if your doubts lead you to read a more advanced text.

To begin with, remember the notation of D for d/dx introduced in section 10.2. Equation (12.6) can now be rewritten as

$$(a\mathrm{D}^2+b\mathrm{D}+c)y=f(x). \tag{12.9}$$

This is the convenient form for setting out to calculate the complementary function and particular integral.

12.4 THE COMPLEMENTARY FUNCTION

The complementary function of the solution to equation (12.6) is, using the form of equation (12.9), the solution to

$$(aD^2+bD+c)y=0.$$

The quantity in the brackets here is a quadratic expression. If D were a variable, one could solve

$$a\mathrm{D}^2+b\mathrm{D}+c=0 \tag{12.10}$$

to yield two roots, which we can call α and β, where α and β are

$$\frac{-b \pm \sqrt{b^2 - 4ac}}{2a}$$

and

$$(\mathrm{D}-\alpha)(\mathrm{D}-\beta)=0.$$

In the case of first-order linear differential equations, an equation could have been written in the style $(\mathrm{D}-\alpha)y=0$ and would have yielded a solution $y=C\,\mathrm{e}^{\alpha x}$. This suggests a solution for a second-order equation which leads to the following rules for obtaining the complementary function.

1. Construct and obtain the roots of the **auxiliary equation**, equation (12.10) above, as if D were a normal variable; call the roots α and β.
2. If $(b^2-4ac)>0$, α and β are real and the complementary function, v, is given by

$$v=A\,\mathrm{e}^{\alpha x}+B\,\mathrm{e}^{\beta x}$$

   where A and B are the arbitrary constants expected in the complementary function.
3. If $(b^2-4ac)<0$, α and β are complex and also are conjugate (cf. Section 7.5), so can be written as $p+iq$ and $p-iq$ (where p and q are real). Then

$$\begin{aligned} v &= \mathrm{e}^{px}(A\cos qx + B\sin qx) \\ &= A'\,\mathrm{e}^{px}\cos(qx+B') \\ &= A''\,\mathrm{e}^{px}\sin(qx+B'') \end{aligned}$$

   where the three versions of v are interconvertible with appropriate choice of the auxiliary angle and arbitrary constants (cf. Section 4.9)
4. If $(b^2-4ac)=0$, the roots both equal a real value α and the complementary function takes the form

$$v=(A+Bx)\,\mathrm{e}^{\alpha x}.$$

Example 12(g) Solve

$$\frac{\mathrm{d}^2y}{\mathrm{d}x^2}+5\frac{\mathrm{d}y}{\mathrm{d}x}+6y=0.$$

Answer: The auxiliary equation is $\mathrm{D}^2+5\mathrm{D}+6=0$, i.e. $(\mathrm{D}+3)(\mathrm{D}+2)=0$, so the roots are -3 and -2. Hence the solution is

$$y=A\,\mathrm{e}^{-3x}+B\,\mathrm{e}^{-2x}.$$

(NB: in this case, and in the next example, the particular integral must be zero since the r.h.s. of the equation is zero.)

Example 12(h) Solve

$$\frac{d^2x}{dt^2} - 4x = 0,$$

given that $x=5$ when $t=0$, and

$$\frac{dx}{dt} \to 0 \text{ as } t \to \infty.$$

Answer: The auxiliary equation is $D^2 - 4 = 0$ so the roots are $+2$ and -2. Hence the solution is

$$x = A\,e^{-2t} + B\,e^{2t}.$$

But $x = A + B = 5$ when $t=0$ and

$$\frac{dx}{dt} = -2A\,e^{-2t} + 2B\,e^{2t}.$$

Since $dx/dt \to 0$ as $t \to \infty$, therefore $B=0$ (otherwise dx/dt would increase as e^{2t} when $t \to \infty$); and so $A=5$ and the solution is $x = 5\,e^{-2t}$.

12.5 THE PARTICULAR INTEGRAL

If $f(x)$ in equation (12.6) is a polynomial, that is, if $f(x)$ can be written as

$$f(x) = p_n x^n + p_{n-1} x^{n-1} + \cdots + p_2 x^2 + p_1 x + p_0$$

then the particular integral, which can be written as y_P, is also a polynomial of the same degree as $f(x)$. This comes about because the l.h.s. of equation (12.6) contains the term cy_P since y_P is a solution of the equation; contribution from the other two terms, derivatives of y_P, will be polynomials of lower degree; and the sum of all three terms must equal $f(x)$. So if

$$y_P = q_n x^n + q_{n-1} x^{n-1} + \cdots + q_2 x^2 + q_1 x + q_0,$$

a series of statements can be made equating the coefficients of each power of x on the two sides. That is, for the coefficients of:

$$\begin{array}{llll} x^n; & cq_n & & = p_n \\ x^{n-1}; & cq_{n-1} + bnq_n & & = p_{n-1} \\ x^{n-2}; & cq_{n-2} + b(n-1)q_{n-1} & + an(n-1)\,p_n & = p_{n-2} \\ & \vdots \quad\quad \vdots & \vdots & \vdots \end{array}$$

until finally,

$$cq_0 \quad + bq_1 \quad\quad + 2aq_2 \quad\quad = p_0.$$

In this way, all the coefficients, q_i, of the particular integral can be successively calculated. Note especially, that if $f(x)$ is just a constant, k, that particular integral is also a constant, with the value k/c.

When $f(x)$ is a more complicated function than a polynomial, finding the particular integral can be very difficult, but there are rules for certain special cases. Since D stands for d/dx, the operator for differentiation, we can guess that 1/D stands for something like integration (not quite, but something like). That's enough of a pointer to the following rules.

1. If, in equation (12.6), $f(x)=\mathrm{e}^{kx}$, to obtain the particular integral, u, divide equation (12.9) through by the bracket, write

$$u=\frac{\mathrm{e}^{kx}}{a\mathrm{D}^2+b\mathrm{D}+C}$$

and then replace D by k, ending up with

$$u=\frac{\mathrm{e}^{kx}}{ak^2+bk+c}.$$

2. If, in equation (12.6), $f(x)=\cos kx$, rearrange as in 1. to get

$$u=\frac{\cos kx}{a\mathrm{D}^2+b\mathrm{D}+c}.$$

Now replace D^2 by $-k^2$ so that

$$u=\frac{\cos kx}{-ak^2+b\mathrm{D}+C}.$$

For convenience write this as

$$u=\frac{\cos kx}{b\mathrm{D}+g},$$

where $g=c-ak^2$.
Rationalize by multiplying top and bottom by $b\mathrm{D}-g$ (cf. Section 5.1) so that

$$u=\frac{(b\mathrm{D}-g)\cos kx}{b^2\mathrm{D}^2-g^2}.$$

In the denominator, replace D^2 by $-k^2$, as before. In the numerator, replace D cos kx by $-k\sin kx$, because D cos kx means (d/dx)(cos kx). Then

$$u=\frac{-bk\sin kx-g\cos kx}{-b^2k^2-g^2}$$

$$=\frac{bk\sin kx+(c-ak^2)\cos kx}{b^2k^2+(c-ak^2)^2}.$$

3. If, in equation (12.6), $f(x)=\sin kx$, then to find the particular integral proceed exactly as in 2, except that in the final step $D \sin kx$ is replaced by $k \cos kx$.

(In these rules I have not mentioned the possibility of the substitutions giving zero in the denominator of u, nor the possibility that either α or β, the roots of the auxiliary equation, equals k; refer to a more advanced text if you meet one of these cases.)

Example 12(i) Solve

$$6\frac{d^2y}{dx^2}+\frac{dy}{dx}-y=e^{-x}.$$

Answer:

(a) *Complementary function.* The auxiliary equation is $6D^2+D-1=0$, i.e. $(3D-1)(2D+1)=0$. The roots are $\frac{1}{3}$ and $-\frac{1}{2}$.
Hence

$$v=A\,e^{x/3}+B\,e^{-x/2}.$$

(b) *Particular integral*

$$u=\frac{e^{-x}}{6D^2+D-1}.$$

Putting -1 for D,

$$u=\frac{e^{-x}}{6-1-1}=\frac{e^{-x}}{4}.$$

(c) The *complete solution* is thus

$$y=A\,e^{x/3}+B\,e^{-x/2}+\frac{e^{-x}}{4}.$$

Example 12(j) Solve

$$\frac{d^2x}{dt^2}+4\omega^2x=3\cos\omega t.$$

Answer:

(a) *Complementary function.* The auxiliary equation is $D^2+4\omega^2=0$. The roots are $\pm 2i\omega$. Therefore

$$v=A\cos 2\omega t+B\sin 2\omega t.$$

(b) *Particular integral*

$$u = \frac{3\cos\omega t}{\mathrm{D}^2 + 4\omega^2}$$

Write $-\omega^2$ for D^2, giving

$$u = \frac{3\cos\omega t}{3\omega^2}.$$

(c) *Complete solution*

$$x = A\cos 2\omega t + B\sin 2\omega t + \frac{\cos\omega t}{\omega^2}.$$

Example 12(k) Solve

$$\frac{\mathrm{d}^2 y}{\mathrm{d}x^2} + 6\frac{\mathrm{d}y}{\mathrm{d}x} + 9y = 3\cos 2x.$$

Answer:

(a) *Complementary function.* The auxiliary equation is $\mathrm{D}^2 + 6\mathrm{D} + 9 = (\mathrm{D}+3)^2 = 0$. Because the roots are equal, the complementary function is $v = (A + Bx)\,\mathrm{e}^{-3x}$.

(b) *Particular integral*

$$u = \frac{3\cos 2x}{\mathrm{D}^2 + 6\mathrm{D} + 9}.$$

Write -4 in place of D^2 to get

$$u = \frac{3\cos 2x}{-4 + 6\mathrm{D} + 9} = \frac{3\cos 2x}{6\mathrm{D} + 5}.$$

Rationalize with $6\mathrm{D} - 5$ to give

$$u = \frac{(6\mathrm{D} - 5)3\cos 2x}{(6\mathrm{D} - 5)(6\mathrm{D} + 5)}.$$

Again write -4 in place of D^2 and write $-2\sin 2x$ in place of $\mathrm{D}\cos 2x$. Hence

$$u = \frac{15\cos 2x + 36\sin 2x}{169}.$$

(c) *Complete solution*

$$y = (A + Bx)\,\mathrm{e}^{-x} + \frac{15\cos 2x + 36\sin 2x}{169}.$$

Example 12(l) The fish in a certain stretch of river are just the two species minnow (*Phoxinus phoxinus*) and perch (*Perca fluviatilis*). The minnows, with population density m per unit length of river, breed at a constant fraction of their number, at a rate $0.7m$ (where this and other rates are given per unit time (a year)), but are limited by food supply and disease and also are preyed on by the perch. The death rate by infection, etc. of the minnows is also proportional to their number, and is $0.5m$. The loss by predation of minnows is practically independent of their number, but is proportional to the number, p, of perch per unit length of river; and it is estimated that each perch consumes 2000 minnows in a year. The perch breed at a rate proportional to their food supply, i.e. proportional to the number of minnows, and the rate is $1 \times 10^{-5}m$. However, from 1 July 1985 (time $t=0$) the river banks were occasionally opened to anglers who thereafter caught perch at a regular cyclical rate which can be approximated by $50(1+\cos 2\pi t)$ per unit length of river, i.e. there was maximum angling activity in midsummer and zero activity in midwinter. When angling began the fish numbers were estimated per unit length of river to be 500 000 minnows and 125 perch. Find how the populations of minnows and perch vary with time in the five years after fishing began, assuming that both species breed continuously through the year (which, of course, is not the case).

Answer: Use the notation $\dot{m}$ for $\mathrm{d}m/\mathrm{d}t$, $\ddot{m}$ for $\mathrm{d}^2m/\mathrm{d}t^2$, etc.
For each species, the rate of increase equals the breeding rate less the rate of loss.
For minnows this can be expressed as

$$\begin{aligned}\dot{m} &= 0.7m - 0.5m - 2000p \\ &= 0.2m - 2000p. \qquad (12.11)\end{aligned}$$

The corresponding equation for perch is

$$\dot{p} = 1 \times 10^{-5}m - 50(1+\cos 2\pi t). \qquad (12.12)$$

Equations (12.11) and (12.12) form a pair of simultaneous differential equations. Such equations generally require further differentiation before they can be solved. From equation (12.11)

$$\ddot{m} = 0.2\dot{m} - 2000\dot{p}$$

and from equation (12.12)

$$\ddot{m} = 0.2\dot{m} - 2000(1 \times 10^{-5}m - 50(1+\cos 2\pi t)).$$

Hence, writing K for 1×10^5, the number of minnows is given by the solution of

$$\ddot{m} - 0.2\dot{m} + 0.02m = K(1+\cos 2\pi t).$$

Complementary function. The auxiliary equation is $D^2-0.2D+0.02=0$. The roots are $(0.2\pm\sqrt{0.2^2-0.08})/2=0.1\pm i0.1$. Write these roots as $\alpha\pm i\beta$ so the complementary function, with arbitrary constants, can be written as $A'\,e^{\alpha+i\beta}+B'\,e^{\alpha-i\beta}$ or as $e^{\alpha}(A'\,e^{i\beta}+B'\,e^{-i\beta})$.

But $e^{i\theta}=\cos\theta+i\sin\theta$, so choosing A' and B' in such a way that $A'=A+iB$ and $B'=A-iB$ (where A and B are two other arbitrary constants) we end up with the complementary function $e^{\alpha}(A\cos\beta+B\sin\beta)$. Thus

$$m_C=e^{0.1t}(A\cos 0.1t+B\sin 0.1t)$$

where A and B still need to be determined.

Particular integrals The l.h.s. of equation (12.13) is in two parts. The particular integral corresponding to the constant term on the l.h.s. of equation (12.13) is also a constant, $m_{P1}=K/0.02=5\times10^6$. The remaining part of the particular integral is

$$m_{P2}=\frac{K\cos 2\pi t}{D^2-0.2D+0.02}.$$

This is dealt with by replacing D^2 in the denominator by $-(2\pi)^2=-39.48$, giving

$$m_{P2}=\frac{K\cos 2\pi t}{-0.2D-39.46}$$

$$=-\frac{(0.2D-39.46)(K\cos 2\pi t)}{(0.2D-39.46)(0.2D+39.46)}.$$

After again replacing D^2 in the denominator by $-4\pi^2$ and using D in the numerator as an operator acting on the cosine function, the particular integral is obtained as

$$m_P=5\times10^6-\frac{K(0.4\pi\sin 2\pi t+39.46\cos 2\pi t)}{1558.55}$$

$$=5\times10^6-80.6\sin 2\pi t-2531.7\cos 2\pi t.$$

Complete solution; determination of constants The complete solution is $m=m_C+m_P$,
i.e.

$$m=e^{t/10}\left(A\cos\left(\frac{t}{10}\right)+B\sin\left(\frac{t}{10}\right)\right)+5\times10^6-80.6\sin 2\pi t-2531.7\cos 2\pi t.$$

Hence

$$\dot{m}=e^{-t/10}\left(\frac{B+A}{10}\cos\left(\frac{t}{10}\right)+\frac{B-A}{10}\sin\left(\frac{t}{10}\right)\right)-2\pi 80.6\cos 2\pi t$$
$$+2\pi 2531.7\sin 2\pi t.$$

But $m=m_0=500\,000$ when $t=0$ so that $A=-4.497\times10^6$. At $t=0$, $(\dot{m})_0=0.2m_0-2000p_0$ from equation (12.11) and since at $t=0$ we have that $p=p_0=125$, it follows that the value of B is 3.003×10^6.

Thus the number of minnows is given by

$$m=5\times10^6+e^{t/10}\left(3.003\times10^6\sin\left(\frac{t}{10}\right)-4.497\times10^6\cos\left(\frac{t}{10}\right)\right)$$
$$-80.62\sin 2\pi t-2.532\times10^3\cos 2\pi t. \qquad (12.14)$$

Substituting equation (12.14) into equation (12.12) and integrating, using the results of example 10(l) and exercises 10E, number 5, gives, after sorting out the arithmetic,

$$p=e^{t/10}\left(-239.9\cos\left(\frac{t}{10}\right)+127.6\sin\left(\frac{t}{10}\right)\right)$$
$$+1.28\times10^{-4}\cos 2\pi t-7.96\sin 2\pi t+C.$$

But $p=p_0=125$ when $t=0$ so that $C=364.9$ and

$$p=364.9+e^{-t/10}\left(127.6\sin\left(\frac{t}{10}\right)-239.9\cos\left(\frac{t}{10}\right)\right)$$
$$+1.28\times10^{-4}\cos 2\pi t-7.96\sin 2\pi t.$$

Graphs of the numbers, m and p, of minnows and perch are shown in Fig. 12.3. (The manner in which the number of minnows is shooting up after five years shows that the ecological conditions stated in the problem have been greatly over-simplified.)

EXERCISES 12C

1. Solve

(a) $\dfrac{d^2y}{dx^2}+4\dfrac{dy}{dx}+3y=0$;

(b) $\dfrac{d^2y}{dx^2}+7\dfrac{dy}{dx}+12y=0$ given that $y=5$ when $x=0$ and $y=1$ when $x=1$.

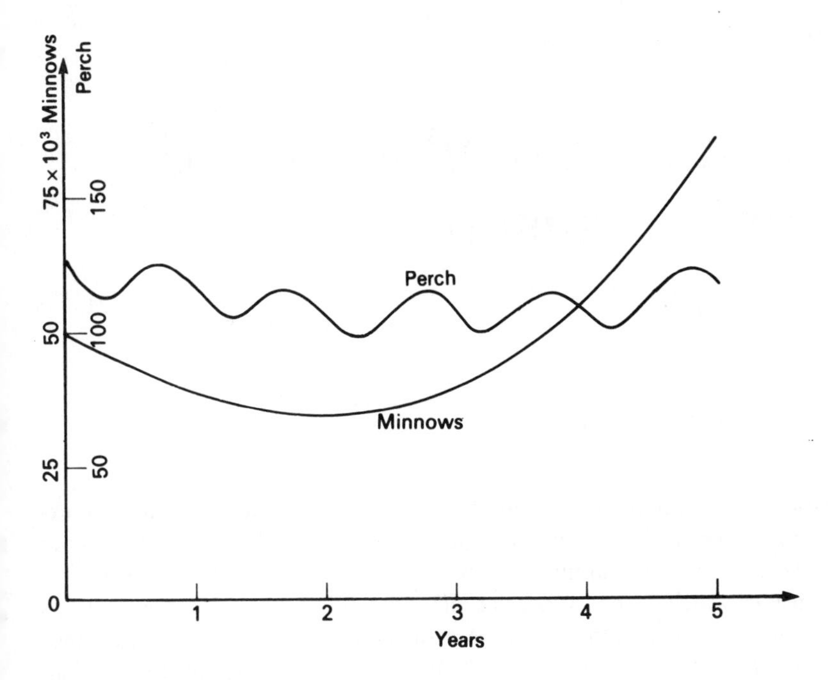

Figure 12.3

2. Solve

$$\frac{d^2y}{dx^2}+3\frac{dy}{dx}+2y=e^{3x}$$

and confirm that you have obtained the correct solution.

3. Solve

$$4\frac{d^2y}{dx^2}+8\frac{dy}{dx}+5y=\sin x$$

and confirm that you have obtained the correct solution.

For Angling may be said to be like the Mathematicks, that it can n'er be fully learnt; at least not so fully, but that there will be more new experiments left for the tryal of other men that succeed us.

Izaak Walton, *The Compleat Angler*

13

Statistics: population and sample measures

Statistics is a branch of mathematics which is not merely useful to experimental science, but is an absolutely essential tool for it. This comes about because, as you surely know from first-hand experience, observation of quantities which do not appear as either small integers or simple fractions are inevitably, and in some cases necessarily, affected by errors; and other observations are of quantities which, by their nature, are variable within limits. The methods of statistics guide the statements which summarize the observations and also circumscribe the inferences which can be drawn from those observations. You will certainly have used many statistical operations on experimental results. The purpose of this chapter and the next is to remind you of them briefly and to give you the opportunity to reconsider the meanings, limitations and implications of the results of these operations.

Statistics is closely associated with the theory of probability. However, I intend to deal peremptorily with that aspect of statistics, avoiding it if possible, or else just quoting the strictly necessary result of the calculus of probability. The three main topics of statistics to be dealt with in this chapter and the following one are: (a) descriptions and measures of data (data distributions); (b) correlations and regression; and (c) hypothesis testing. There are, however, a few preliminaries of which you should remind yourself.

13.1 DISTRIBUTIONS OF DATA, CONTINUOUS AND DISCONTINUOUS

A recorded observation or datum is called a statistic; the property being recorded, when it is used mathematically, is called a variate; and a number of observations constitutes a sample, taken from a population (or sometimes said to be taken from a universe). What distinguishes a variate from a variable is that the former has a probability associated with it, related to the population from which the statistic and the sample are drawn, and sometimes relating to the manner of the drawing; but that matter is not to be pursued here.

The sample is necessarily of finite size, but the population may be finite or infinite; and if the population is finite, the sample may be complete (encompassing 100% of the population) or partial. As an example, an electoral roll records the total population eligible to vote, an opinion poll records the opinions of a sample, usually small, of the electorate and a general election provides a large sample of electoral opinion, which sometimes, in some countries, is claimed to be complete. The records of observations may be of a variate which is discrete (and often integer), like the number of children to which members of a sample of women have given birth, or which is continuous, like the size of each woman. Furthermore, the observations may be grouped, especially when the variate is continuous so that when the records of observations are graphed, a histogram results and, if this is translated into manufacturing practice, the women who don't make their own clothes are forced into sizes ..., 10, 12, 14, 16, ..., or ..., 38, 40, 42, 44, ..., depending on the country they live in, and the individuality of their sizes is lost.

The relation between number of statistics and value of variate is the distribution of the variate. If the variate takes only discrete values a graph of the distribution is necessarily discontinuous and forms a histogram. If the population is infinite (really so or effectively so) the use of relative frequency of occurrence rather than absolute number of instances is unavoidable and if the variate is continuous the distribution is a smooth curve. But sometimes it is convenient for histograms to be smoothed out. For example, the amount, A, of radioactive material left after time t is always described by a formula like $A = A_0\,e^{-0.693t/t_{\frac{1}{2}}}$, even though radioactivity is a discrete event, happening atom by atom. Alternatively, it may be convenient for smooth curves to be cut into strips and presented as histograms.

Because you have almost certainly met all these ideas before, the various matters which follow can be discussed in relation to distributions which are not simple, but which are typical of the awkwardness of the real world. Table. 13.1 and Fig. 13.1 relate to the brightness of the electric lights in the house I live in and Table 13.2 and Fig 13.2 relate to the brightness of stars in the sky. The sample of Table 13.1 is complete in relation to this house, but is only about 1% in relation to the street in which I live and is something like 0.000 05% in relation to the city of London. This sample is certainly discrete and finite in respect of the number of lights and also the manufacturer's power rating on each one, which allots them into only five groups (bins) although, as a matter of direct experience, the lights in any one group are not equally bright. Table 13.2, in contrast, is not the record of a

Table 13.1 Electric light ratings in my house

	Rating, x_i (watts, W)	Number, n_i	$n_i x_i$
	20	3	60
	40	10	400
	60	12	720
	100	17	1700
	150	1	150
Totals:		43	3030

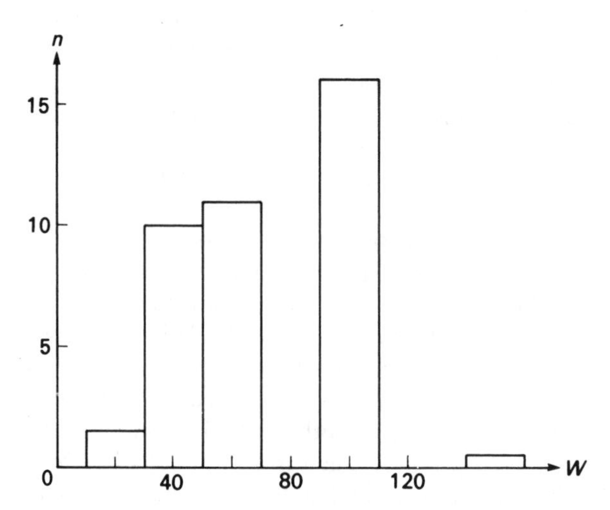

Figure 13.1

sample, but is an estimate obtained from a number of small samples of various parts of the sky, extrapolated to cover the whole sky. The numbers are, in principle, discrete but after the first few entries in the table they are so large that they can be taken as if coming from an infinite population. Equally, the variate, stellar visual magnitude, which depends on a star's luminosity and distance, is continuous, so a smooth graph of the distribution, like that of Fig. 13.2, is in order. However, it is interesting to note in respect of Fig. 13.2 that although it many be possible to take the graph further to the right with more powerful telescopes, we just don't know where the end of the universe happens, so we don't know if or where the graph might stop.

Table 13.2 Visual magnitudes of stars

Magnitude	Number, n_i^*	$\log n_i$	Magnitude	Number, n_i^*	$\log n_i$
0	3	0.40	10	207 000	5.32
1	9	0.95	11	546 000	5.74
2	29	1.46	12	1 400 000	6.15
3	97	1.99	13	3 430 000	6.54
4	392	2.59	14	8 100 000	6.91
5	1 090	3.09	15	18 200 000	7.26
6	3 230	3.55	16	39 000 000	7.59
7	9 450	3.98	17	79 000 000	7.90
8	26 700	4.43	18	146 000 000	8.16
9	76 000	4.88	19	264 000 000	8.42
			20	440 000 000	8.64

*Numbers against magnitude m are in the range $m-0.5$ to $m+0.5$.
Stellar brightness decreases as magnitude increases.

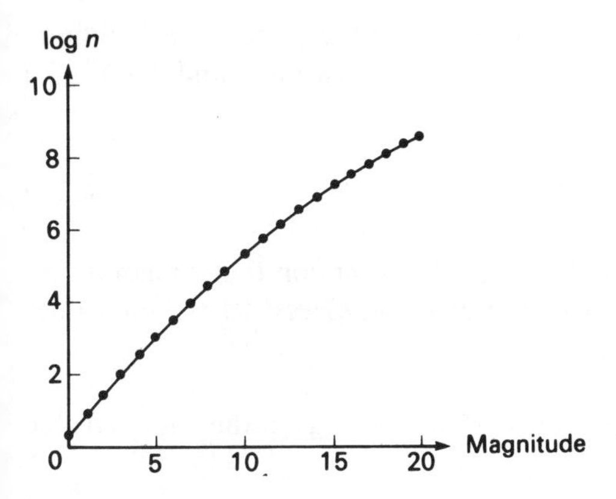

Figure 13.2

13.2 MEASURES OF LOCATION: MEAN, MEDIAN AND MODE

A distribution is described by its position or location on the abscissa axis and by its shape. Position is particularly important and the most frequently used and best-known measure of position is the **mean**. The mean, m, is defined for a sample of n observations x_i (where $i=1, 2, 3, \ldots n$) by

$$m = \frac{\sum_{i=1}^{m} x_i}{n}.$$

In the case where observations have been grouped so that there are n_i with value x_i, n_j with value x_j, etc., $\sum_i n_i = n$, so we have

$$m = \frac{\sum_i n_i x_i}{n} = \sum_i \frac{n_i}{n} x_i.$$

Sometimes it is convenient to write $n_i/n = f_i$, where f_i is the fraction or frequency of appearance of x_i. We then have

$$m = \sum f_i x_i. \tag{13.1}$$

For a population of ν individuals, with values ξ_i of the variate in question, the mean, μ, is given by

$$\mu = \frac{\sum \xi_i}{\nu} = \frac{\sum \nu_i \xi_i}{\nu} = \sum \phi_i \xi_i \tag{13.2}$$

if the population is finite and putting $\phi = \nu_i/\nu$. If the population is infinite and a fraction $\phi(\xi)$ has the variate value lying between ξ and $\xi + \delta\xi$, the mean is

$$\mu = \int \phi(\xi)\xi \, d\xi. \tag{13.3}$$

(NB: use is made here of the widely accepted convention that roman letters are used for values associated with samples and Greek letters for values associated with the population.)

Example 13(a) What is the mean value of light ratings in the data of Table 13.1?

Answer: $n=43$, $\sum n_i x_i = 3030$ W; hence $m = \frac{3030}{43}$ W $= 70.5$ W.

The answer to example 13(a) illustrates one of the objections to the use of the mean as a measure of location. No one manufactures lights with a rating of 70 W, so the conclusion that this is the average value is of little help in designing interior lighting levels or in planning light stocks either in the domestic store cupboard or on the supermarket shelf. The most used alternative measure is the **median**. To obtain the median the data must be

Table 13.3 Examination marks of students

Place number	Mark	Place number	Mark
1	63	11	57
2	62	12	56
3	62	13	56
4	60	14	48
5	59	15	42
6	59	16	30
7	59	17	24
8	59	18	18
9	58	19	15
10	58	20	8

arranged in rank order, as in Tables 13.1–13.3. Then the median value of the distribution is the middle statistic in the rank. In Table 13.1, with $n = 43$, the middle statistic is number 22, which is one of the last of the 60 W lamps.

The median of a distribution is the value of which one can say that, within the sample, there are as many values less than or equal to it as there are greater than or equal to it. The mean can be far from the centre of the ranked order of members of the sample. For example, Table 13.3 gives the examination marks (not untypical) of a class of students. The average or mean mark is 47.6 so that nearly three quarters of the students (14 out of 20) are above average. The median mark is 57.5 (midway between the tenth and eleventh marks in the ranking order); half of the class have a higher mark than the median and half a lower mark. However, one difficulty about using the median is that the observations must be ranked; this is typically done when results are presented, as in Tables 13.1–13.3, but to do so during the period of data collection may be difficult, whereas it is easy to keep track of the mean during this time.

It can be seen that cases can be made for either the mean or the median as the appropriate measure of position of a distribution, but when a distribution is very asymmetric, like that of star magnitudes in Table 13.2, neither is very helpful. The mean stellar magnitude according to Table 13.2 in 18.87 and the median is 19.27, both values coming very near the end of the distribution, because there are such vast number of very dim stars.

A third measure of position of a distribution, occasionally mentioned but rarely used, is the **mode**. This is the value of statistic which appears with the greatest frequency. Thus the mode of light rating (Table 13.1) is 100 W, the mode of stellar magnitude (Table 13.2) is 20, and the mode of examination marks (Table 13.3) is 59. The mode is the position of a peak in a graph of distribution (e.g. the 100 W position in Fig. 13.1) but its only real use is, with a slight extension of its definition, to draw attention to the existence of more than one peak in a distribution.

Table 13.2 is scientifically curious in that, with the naked eye, stars are visible down only to magnitude 6 or 7. For stars of higher magnitude to be seen, a telescope must be used together with detecting instrumentation for the very highest magnitudes. But 'visual magnitude' remains a proper astronomical term as distinct from 'photographic magnitude'. Stars appear in many colours and the eye and photographic film have distinctly different colour sensitivities. So the entries in Table 13.2 mix together many different kinds of stars (i.e. types *O*, *B*, *A*, *F*,...) which properly should be considered separately. There is the same problem within Table 13.1: fluorescent tubes have been entered by their rating in watts, even though they are far more luminous than filament bulbs. Talbe 13.1 reveals information about average energy dissipation rather than anything about the brightness of my rooms.

The purpose of this long digression on measures of location is to stress than you should treat with caution any statement purporting to have statistical meaning until you are satisfied that the data have been properly gathered and appropriately analysed. (I have omitted to stress that correct units must be used in an analysis; for example, it would be wrong to seek the mean stellar magnitude by basing an average on the logarithms of the numbers, which have been introduced only to allow a reasonable plot of the data, as in Fig. 13.2.)

13.3 DISTRIBUTION MOMENTS

From equation (13.2) we can write

$$\sum \phi_i(\xi_i - \mu) = 0. \tag{13.4}$$

This equation follows because the l.h.s of equation (13.4) can be written as $\sum\phi_i\xi_i - \sum\phi_i\mu$ which in turn can be written as $\sum\phi_i\xi_i - \mu\sum\phi_i$ since μ is a constant. But $\sum\phi_i\xi_i = \mu$ from equation (13.3) and $\sum\phi_i = \sum v_i/v = 1$; hence equation (13.4) follows. The l.h.s. of equation (13.4) is called the first moment about the mean of the distribution. Compare this to a mechanical model in which a number of 1 kg weights are placed at integer positions on a balanced rod. The distribution is given in Table 13.4.

Table 13.4 Weights on a rod

Position	1	2	3	4	5	6	7
Weight (kg)	2	4	7	2	6	6	1

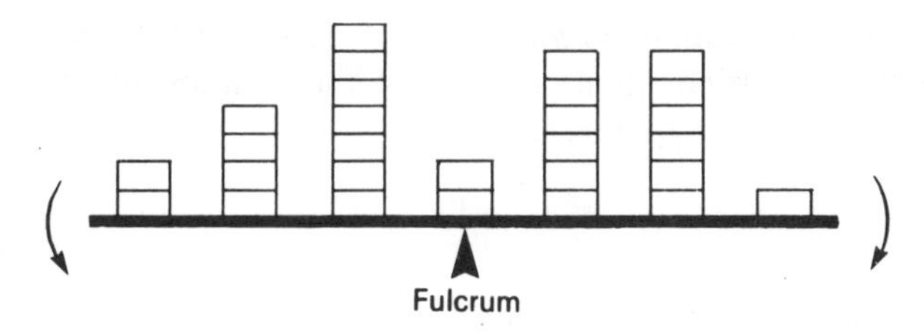

Figure 13.3

As illustrated in Fig 13.3, the weights impose torques or turning moments about the fulcrum, where such a torque or moment is defined as equal to (force) × (distance from fulcrum). The mean of the distribution is at the position

$$\frac{2\times1+4\times2+7\times3+2\times4+6\times5+6\times6+1\times7}{2+4+7+2+6+6+1}=4.$$

If the fulcrum is placed at position 4, the clockwise moment, from the weights at positions 5, 6, and 7 is $(6\times1+6\times2+1\times3)g=21\,g$ (where g is the acceleration due to gravity), and the anticlockwise moment is $(2\times3+4\times2+7\times1)g=21\,g$, from the weights at positions 1, 2, and 3. Thus the two moments just cancel and the rod balances; or to put it another way, the total moment about the mean (the point of balance) is zero.

In fact, equation (13.4) is merely the first of a family of equations which in general are written as

$$\mu_r=\sum\phi_i(\xi_i-\mu)^r \tag{13.5}$$

where μ_r is called the rth moment about the mean of the distribution. The moments about the mean are used to describe the shapes of distributions. It is because the mean plays a central role in the moments of the distribution that it is especially favoured over the median as a descriptor of a distribution.

13.4 POPULATION MEASURES OF DISTRIBUTION SHAPE

The rth moment about the mean, given by equation (13.5), is for the distribution of a discrete variate. To include the case of continuous variates, we should write

$$\mu_r=\sum\phi_i(\xi_i)(\xi_i-\mu)^r \qquad \text{or} \qquad \int\phi(\xi)(\xi-\mu)^r\,d\xi.$$

However, for simplicity, the following account is given in terms of a discrete variate.

The first moment about the mean, μ_1, is zero. In effect, its significance has been used up in the determination of the mean and the location of the distribution, of which sufficient has been said above.

Variance

The second moment about the mean of a distribution, μ_2, is called the variance. It is used as a measure of the breadth or dispersion of a distribution. Most often, however, the square root of the variance, called the standard deviation (defined below), is used for this purpose.

Skewness

The third moment about the mean, μ_3, has no special name, but is a measure of the extent to which the dispersion is skewed or non-symmetric. If $\mu_3 = 0$, the distribution is symmetric about the mean. If $\mu_3 > 0$, the distribution is extended (has a tail) on the side where the variate is greater than the mean (which is usually, but not always, the right-hand side). If $\mu_3 < 0$, the distribution is extended on the side where the variate is less than the mean, usually the left-hand side as illustrated in Fig. 13.4. Since μ_3 has the dimension of the variate cubed, it is sometimes replaced as a descriptor of skewness by a non-dimensional quantity denoted by β_1 where

$$\beta_1 = \frac{\mu_3}{(\mu_2)^{\frac{3}{2}}}.$$

Kurtosis

The kurtosis of a distribution curve is its property of being excessively

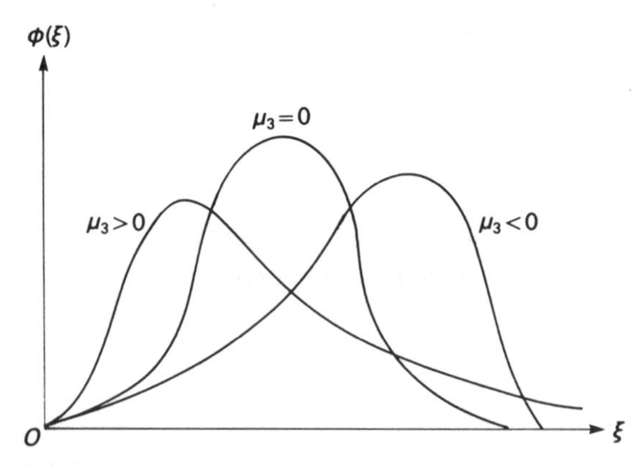

Figure 13.4

peaked or excessively flattened. Comparison is with the standardized Gaussian or normal curve,

$$\phi(\xi)=\frac{1}{\sqrt{2\pi}}e^{-\frac{1}{2}\xi^2} \tag{13.6}$$

as shown in Fig. 13.5. The fourth moment about the mean is a measure of kurtosis, although usually after it has been corrected for dimensionality. This is done by introducing a non-dimensional quantity β_2, where

$$\beta_2=\frac{\mu_4}{(\mu_2)^2}.$$

Then for the normal curve, $\beta_2=3$; if $\beta_2>3$, the distribution is excessively peaked at the centre; if $\beta_2<3$, the curve is flattened.

Standard deviation

The variance is a measure which depends on the square of the variate but for most purposes it is more convenient to express the spread of a distribution in terms of its mean and some quantity with the same dimension as the mean. For this purpose, the standard deviation, σ, is now invariably used, where $\sigma=\sqrt{\mu_2}$. As well as being used it is frequently abused. It is a very useful measure for symmetric curves which are near to the normal Gaussian shape of equation (13.6) and the curve marked $\beta=3$ in Fig. 13.5; and that can be of assistance in testing whether two samples may have been drawn from the same population. But it can be deceptive when applied to distributions which are very skewed like those of Tables 13.1 and 13.2.

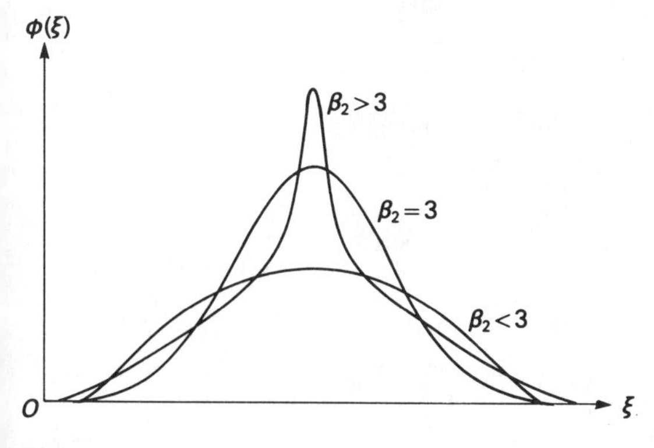

Figure 13.5

13.5 MEASURES OF DISTRIBUTION SHAPE FOR SAMPLES

The various measures just listed were written with Greek letters and were implicitly for infinite populations. For application to finite samples the formulae need to be changed slightly. Also it is worth mentioning various other empirical descriptors of distributions.

The moments about the mean of a distribution involve the frequency variable ϕ. In a sample, as shown in equation (13.1), the frequency is the quantity $f_i = n_i/n$ where n_i is the number of instances of the variate being observed with the value x_i. A full analysis of the calculation of moments about the mean for samples shows that, in determining the mean, m, some of the information inherent in the sample is used up. This is expressed by saying that since one quantity, the mean, has been determined, the data of the sample have lost one degree of freedom, so instead of there being n independent items of data in the sample, it is as if, once m has been specified, there were only $n-1$ independent items. The effect of this analysis is that to obtain many (though not all) of the *sample* formulae it is necessary to multiply the corresponding *population* formula by the factor $n/(n-1)$. Hence, for example, the rth moment about the mean of a sample is given by

$$m_r = \sum n_i \frac{(x_i - m)^r}{n-1}.$$

13.6 VARIANCE AND STANDARD DEVIATION

The variance, v, of a sample in which $n_i = 1$ for all i is given by

$$v = \frac{\sum (x_i - m)^2}{n-1}. \tag{13.7}$$

Multiplying out the numerator of the r.h.s. gives

$$\begin{aligned}\sum (x_i - m)^2 &= \sum (x_i^2 - 2x_i m + m^2) \\ &= \sum x_i^2 - 2m \sum x_i + nm^2 \\ &= \sum x_i^2 - \frac{\left(\sum x_i\right)^2}{n} \qquad \text{since } m = \frac{\sum x_i}{n},\end{aligned}$$

so an alternative formula for the variance is

$$v = \frac{n \sum x_i^2 - \left(\sum x_i\right)^2}{n(n-1)}. \tag{13.8}$$

The standard deviation, s, of a sample is given by the square root of equation (13.7) or equation (13.8):

$$s=\sqrt{v}=\sqrt{\frac{\sum(x_i-m)^2}{n-1}}=\sqrt{\frac{n\sum x_i^2-\left(\sum x_i\right)^2}{n(n-1)}}. \tag{13.9}$$

Many electronic calculators, claiming to offer the function of calculating the sample standard deviation, in fact calculate the population deviation; that is, the calculation is made with n in place of $n-1$ in the denominator of equation (13.9). This error makes a difference of about 5% if $n=10$, of about 2% if $n=25$, of about 0.5% if $n=100$, and becomes very small indeed as n becomes very large. Whether or not such an error is of significance for you depends on the nature of your work, but you should be aware of it.

13.7 SKEWNESS AND KURTOSIS

The formulae for skewness and kurtosis correspond to those for populations, but with the correction for n, i.e. for degrees of freedom, viz.

$$m_3=\frac{\sum(x_i-m)^3}{n-1} \quad \text{and} \quad \beta_1=\frac{m_3}{s^3},$$

$$m_4=\frac{\sum(x_i-m)^4}{n-1} \quad \text{and} \quad \beta_2=\frac{m_4}{s^4}.$$

Skewness is also sometimes measured by the Pearson coefficient, κ, where

$$\kappa=\frac{3(\text{mean}-\text{median})}{s}.$$

13.8 COEFFICIENT OF VARIATION

A frequently used measure of the breadth, or the dispersion, of a sample is its coefficient of variation, CV, defined as

$$\text{CV}=\frac{100s}{m}.$$

That is, the coefficient of variation is the standard deviation expressed as a percentage of the mean. The problem with this measure is that its value depends so crucially on the value, m, of the mean. If the variate can take only positive values then CV usually gives a good idea of the sharpness or otherwise of the distribution. But, if the variate can be either positive or negative, a sample may have a mean near to zero and then have a very large value of CV, despite also having a small standard deviation.

Example 13(b) Calculate the standard deviation of the distribution of data in Table 13.1.

Answer: Since the data is grouped and the mean has been calculated in example 13(a), it is simplest to make the calculation as

$$s=\sqrt{\frac{\sum n_i(x_i-m)^2}{n-1}}.$$

Hence

$$\begin{aligned}s&=((3\cdot(50.5)^2+10\cdot(30.5)^2+12\cdot(10.5)^2\\&\quad+17\cdot(29.5)^2+1\cdot(79.5)^2)/42)^{\frac{1}{2}}\\&=30.6.\end{aligned}$$

EXERCISES 13A

1. Use an electronic calculator to find the standard deviation of the distribution of data in Table 13.4. If the calculator has a standard deviation (s.d.) function, check whether it uses the correct formula or the erroneous one.
2. Calculate the third moment about the mean for the data of Table 13.1.

13.9 QUARTILES, PERCENTILES, ETC.

Corresponding to the division of the sample into halves by the median, there are further divisions which can be used empirically as descriptors of distribution shape. The best established of these are the quartiles of which the first or upper, q_1, divides the ranked data at the statistic a quarter of the way down the rank, the second, q_2, is the median, and the third, or lower quartile, q_3, is at the statistic three-quarters of the way down the rank as shown in Fig. 13.6. The numerical difference between the first and third quartiles is then a measure of the breadth of the distribution, and the difference between (q_2-q_1) and (q_3-q_2), or their ratio, can give a measure of the skewness.

The procedure of finding quartiles has been extended in an obvious way to finer divisions, though with some linguistic difficulties. Octiles divide the distribution into eighths; deciles divide it into tenths. Percentiles divide it into whatever percentage may be specified: e.g. the sixth percentile is a

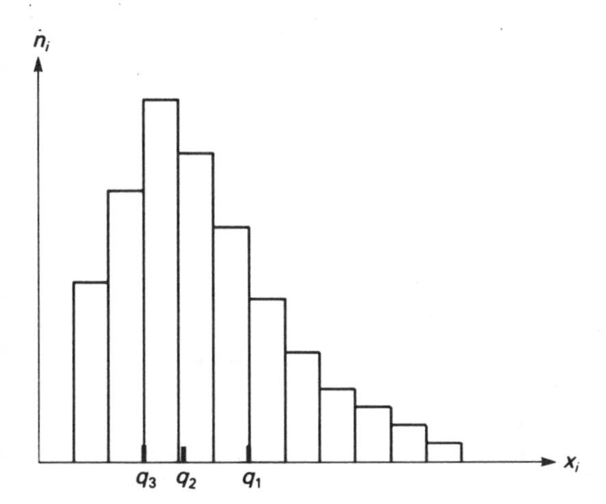

Figure 13.6

division of the ranked order 6% of the way down it. However, there is no need to pursue this fractionation further.

Example 13(c) Find the first and third quartiles of the distribution of the data in Table 13.3.

Answer: The first quartile, q_1, is between places 5 and 6; hence $q_1 = 59$. The third quartile is between places 15 and 16, hence $q_3 = 36$. Note that the median is 58.5 and the skewness of the distribution is indicated by the difference between $(59 - 58.5) = 0.5$ and $(58.5 - 36) = 22.5$.

EXERCISE 13B

Find the first and third quartiles of the distribution of the data in Table 13.1.

13.10 FOUR IMPORTANT DISTRIBUTIONS

Data collected experimentally are found to be distributed in a very large number of ways, but there are four distributions which are found particularly often in the sciences. These are sketched in Fig. 13.7 and are, respectively, (a) the rectangular or flat distribution, (b) the Gaussian or normal, (c) the binomial, and (d) the Poisson. They are described very briefly here as reminders, in the expectation that you have met them before. In describing these distributions, the account is of what is called the probability density function ϕ which, operating on the variate x, gives the fraction of the population expected to have that value of the variate. Because all such

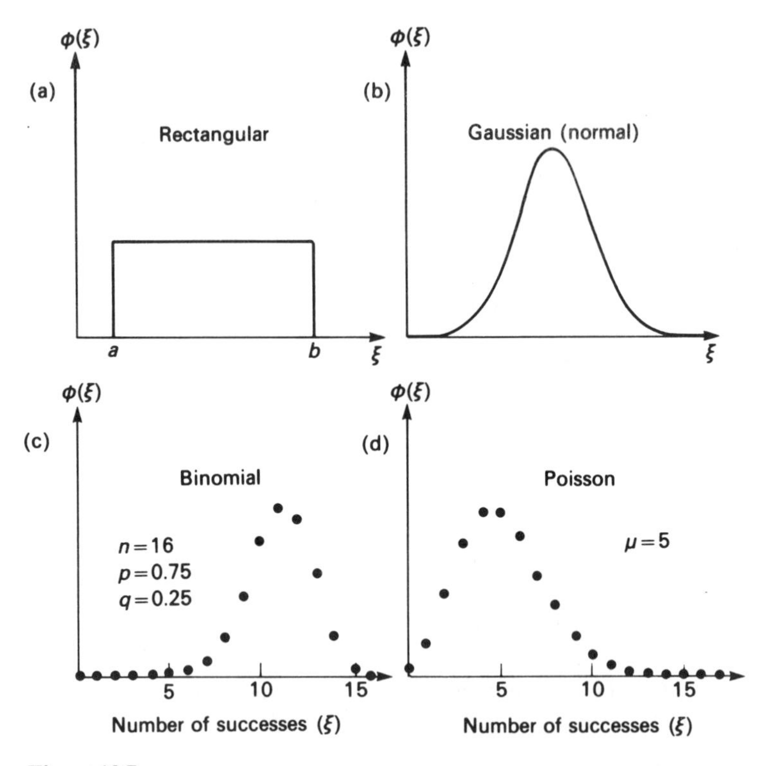

Figure 13.7

fractions must add up to the whole population, it follows that

$$\int \phi(x)\,\mathrm{d}x = 1 \text{ or } \sum \phi(x_i) = 1,$$

depending on whether the distribution is continuous or discrete.

The rectangular or flat distribution

If an inexperienced darts player throws a dart at a board, aiming for the bull's eye, a miss is the most probable outcome and a line joining the centre of the board to the position where the dart strikes has equal probability of being at any angle whatever to the vertical. If an inexperienced croupier throws a ball into a spinning roulette wheel, the ball has equal probability of coming to rest in any one of the 37 slots in the wheel. These are two examples of rectangular or flat probability distributions, the first being continuous and the second discrete.

In the case of the thrown dart, the probability distribution function is

$\phi(\theta)=1/2\pi$ so that $\int_0^{2\pi}\phi(\theta)\,d\theta=1$ as required, when the integration is made over all possible values of the angle θ. In the case of the roulette wheel the probability of each final position of the ball should be 1/37 so that $\Sigma_{i=1}^{37}\phi(x_i)=1$ where x_i is the slot in which the ball lands. (It is said that experienced croupiers can throw a ball on to the roulette wheel with a good expectation of where it will non-randomly land but I have not been able to obtain confirmation of this.)

Generally, if a rectangular distribution extends between the values p and q, the function ϕ for a continuous distribution is $1/(q-p)$ for values of the variate between those limits, and zero for all other values. For a discrete distribution the value of ϕ depends on how many possible values the variate x can take between the two extreme values, but again it is zero outside that range. The mean, μ, in either case, is $(p+q)/2$ and the standard deviation, σ, is $(q-p)/2\sqrt{2}$.

The Gaussian or normal distribution

The so-called standardized form of this distribution is given by equation (13.6); in this case the variate has a mean value of 0 and a standard deviation of 1. For the more general situation of a mean value μ and a standard deviation σ the function ϕ has the form

$$\phi(x)=\frac{1}{\sigma\sqrt{2\pi}}e^{-(x-\mu)^2/2\sigma^2}. \tag{13.10}$$

It can be shown that the requirement that $\int_{-\infty}^{\infty}\phi(x)\,dx=1$ is met. The integral covers all the values which the variate might in principle take, from $-\infty$ to $+\infty$; and the meaning of the integral is that the sum of all possible probabilities is the certainty that something – the particular observations which have been found – happened. For a particular value of the variate, the probability of its lying in the range between x_1 and $x_1+\delta x$ is $\phi(x_1)\delta x$.

Tables for the Gaussian distribution invariably relate to the standardized form of equation (13.6). Sometimes the tables give ϕ for values of x. More often, however, they list values of $\int_{-\infty}^{z}\phi(x)\,dx$ which give the probability of finding a value of the variate less than z; or values of $\int_z^{\infty}\phi(x)\,dx$ which give the probability of finding a value of the variate greater than z; or else the integration limits are 0 and z. However, these various values are interdependent and can often be inferred one from another.

A reasonable argument can be made that, if measurements of a quantity are subject to many independent sources of error and if the quantity has a true value μ, then repeated measurement will show a distribution of results like that of equation (13.10), provided that the errors are random (i.e. not biased in one direction) and that the chance of making a very large error diminishes sensibly with the size of the error. Because it is supposed in laboratory environments that this is the usual state of affairs, the distribution, which was first fully studied by Gauss in the eighteenth century, is called the normal distribution. It has the property that 68.3% of the data (i.e. just over two thirds) will fall in the range $\mu-\sigma$ to $\mu+\sigma$ and 95.5% (i.e. more than nineteen twentieths) in the range $\mu-2\sigma$ to $\mu+2\sigma$.

These statements are made for the distribution of an infinite population, i.e. an infinite number of normal measurements. A finite sample in the real world should be scrutinized to see if it does or doesn't look as if it follows a Gaussian distribution pattern. A formal examination of this question involves hypothesis testing (i.e. testing the hypothesis that the data is indeed normally distributed) which we meet in the next chapter.

The binomial distribution

If a test or trial has only two outcomes, say P and Q, which might be success or failure, up or down, on or off, etc. there may be a definite constant probability attached to each outcome. Suppose the probability of P is p and of Q is q, so that $p+q=1$ (since no other outcome is possible). Then if the test is repeated n times, and if the repetitions are independent (i.e. do not influence one another) the probability of r outcomes P and s outcomes Q, where $r+s=n$, is the coefficient of P^rQ^s in $(pP+qQ)^n$. The binomial expansion (equation 3.5)) shows that this coefficient is ${}^nC_rp^rq^s={}^nC_rp^rq^{n-r}$.

In using the binomial distribution, P and Q are supposed to be understood and the distribution is written as

$$\phi(r)={}^nC_rp^rq^{n-r} \tag{13.11}$$

where the variate can now take only integer values (for which reason r is written, rather than x) between 0 and n. For this distribution, ϕ is necessarily a discrete function. $\sum_{r=0}^{n}\phi(r)=1$ because $\sum_{r=0}^{n}\phi(r)=(p+q)^n=1$ since $p+q=1$. If P is called a 'success' and Q a 'failure', the binomial distribution gives the mean number of successes as np and the standard deviation as $\sqrt{npq}$.

The Poisson distribution

In applications governed by the binomial distribution, p and q are usually of similar value, differing possibly by a factor of up to 10, but rarely by much more. However, if p falls to very small values yet remains proportional to $1/n$ as n becomes very large, so that np remains constant at a value μ, then the binomial distribution becomes modified into the Poisson distribution for which the formula is

$$\phi(r)=\frac{e^{-\mu}\mu^r}{r!} \tag{13.12}$$

where the variate r takes only positive integer values. Here, too, the distribution is necessarily discrete and $\sum_{r=0}^{\infty}\phi(r)=1$ because

$$\begin{aligned}\sum_{r=0}^{\infty}\phi(r)&=e^{-\mu}\sum\frac{\mu^r}{r!}\\&=e^{-\mu}\left(1+\frac{\mu}{1}+\frac{\mu^2}{2!}+\frac{\mu^3}{3!}+\cdots\right)\\&=e^{-\mu}e^{\mu}=1 \qquad \text{by equation (3.7).}\end{aligned}$$

For the Poisson distribution, both the mean and the variance have the value μ, so that the standard deviation is $\sigma=\sqrt{\mu}$.

13.11 THE CALCULATION OF OUTCOME PROBABILITIES

If the appropriate parameter or parameters of a sample are known, such as the mean and standard deviation, they are the best estimates of the corresponding population parameters in the absence of any other information. When the context allows it, a theoretical distribution, such as the binomial, Poisson or normal distribution, may be inferred. It is then possible to calculate, by use of one of equations (13.10)–(13.12), the frequency of a particular value of the variate. The frequency is the fraction of trials which yield that value as outcome and so is the probability of the outcome.

Example 13(d) A medical practitioner finds that, for 40% of his patients, the conditions with which they present at his surgery are the result of poor diet. What is the probability that, on a particular morning, (a) eight out of ten and (b) two out of ten patients in his surgery will be suffering from conditions due to this cause?

Answer: Call a diagnosis of poor diet a 'success'. Hence the frequency distribution of r successes in 10 trials is given by the binomial distribution as the term containing p^r in $(p+q)^{10}$, namely,

$$\phi(r)={}^{10}C_r p^r q^{10-r}.$$

It is given that $p=0.4$; hence $q=0.6$.
(a) For $r=8$, $\phi(8)={}^{10}C_8(0.4)^8(0.6)^2=0.0106$.
(b) For $r=2$, $\phi(2)={}^{10}C_2(0.4)^2(0.6)^8=0.1209$.

Example 13(e) The number of deaths per week at a large (600 bed) general hospital has a mean of 4. What is the probability of (a) just one death in a week, (b) more than six deaths in a week?

Answer: Since the probability of death per patient is very low, assume that the deaths are Poisson distributed. If $\mu=4$ in the Poisson distribution, the expected frequency of r deaths is given by

$$\phi(r)=\frac{e^{-4}4^r}{r!}.$$

Thus the expected distribution is:

r:	0	1	2	3	4	5	6	7
$\phi(r)$:	0.018	0.073	0.146	0.195	0.195	0.156	0.104	0.06

Hence the probability of one death in a week is 0.073, i.e. about 1 in 13.5. The sum of all frequencies is 1.000. The sum of frequencies up to and including $r=6$ is 0.889. Hence the sum of all frequencies for $r>6$ is 0.111, i.e. about one in nine.

Example 13(f) The numbers of men's shoes sold are normally distributed about a mean size of 10 with a standard deviation of 1.0. If a shoe store expects to sell 100 pairs of size 10 shoes in a week, how many of size $11\frac{1}{2}$ is it likely to sell?

Answer: The distribution is stated to be normal with $\mu=10$ and $\sigma=1.0$ so the frequency distribution is

$$\phi(x)=\frac{1}{\sqrt{2\pi}}e^{-(x-10)^2/2}.$$

However, this equation is continuous and shoe sizes vary discontinuously in units of 0.5. Hence the probability density P_S of size S is $\int_{S-\frac{1}{4}}^{S+\frac{1}{4}}\phi(x)\,dx$. Tables of integrals of the Gaussian function then give the following values:

$$\int_{9.75}^{10.25}\phi(x)\,dx=P_{10}=0.5987-0.4013=0.1974$$

$$\int_{11.25}^{11.75}\phi(x)\,dx=P_{11.5}=0.9332-0.9115=0.0217.$$

Hence the expected sale of size $11\frac{1}{2}$ shoes is $(0.0217/0.1974)\times 100$. Thus, the expected sale is 10.99, i.e. approximately 11 pairs.

EXERCISES 13C

1. A large number of trials establish that a dose, LD_{50}, of an insecticide kills 50% of the fly population to which it is administered, with a normal distribution of standard deviation 20%. What is the probability that another trial with the same dose, delivered under identical conditions, results in not more than 30% of flies being killed?
2. Monocytes comprise 5% of white cells in normal blood. Assuming a Poisson distribution of monocyte count, what is the probability of a diagnosis of monocytosis (8% or more monocytes) from a count of 100 white cells made on a blood sample given by a normal healthy subject?
3. If one in ten people entering the country engages in minor smuggling (e.g. bringing in an extra bottle of whisky) what is the chance that two or more out of five people stopped by customs officers will be carrying contraband?

13.12 FITTING EXPERIMENTAL DATA TO A DISTRIBURTION

Suppose that an experimental sample consists of n statistics in groups such that there are n_1 of value x_1, n_2 of value x_2, and so on. Write $n_i/n = f_i$, the frequency of occurrence of x_i. The mean, m, and the standard deviation, s, can be calculated by use of equation (13.1) and equation (13.9). The values m and s can then be used to calculate what the results would have been had they fallen in accord with a known distribution, D.

Of course, making this calculation doesn't in any way make the sample actually follow the distribution D. On the contrary, the scientific procedure is to say: (a) here is the actual sample; and (b) here is the ideal sample which would have been expected if D were to determine the observed values. Then the crucial question of science can be asked: what odds can be given on the truth of the hypothesis that the actual sample was indeed determined by D? Sometimes the fit between (a) and (b) is so close – as when Newton compared the motions of the planets with the predictions of his theory of gravitation, or when Mendel compared the distribution of his varieties of peas with that predicted by his theory of inherited characters – that the sizing of odds is forgotten and the scientist just says 'That's it'. Quite often, too, scientists have said 'That's it' without sufficient justification, when they should have known better. What they should have known about is the procedure of hypothesis testing, which is briefly sketched in Chapter 14.

Here we merely look at the task of fitting data to the distributions given above – the normal, binomial and Poisson. In tackling this task the first requirement, as always, is to apply common sense and as much insight as possible to the problem. The binomial distribution comes about as the result of processes which are either/or, success/failure, win/lose, live/die, with the probabilities remaining constant. The Poisson distribution is similarly limited, typically to events which did or didn't happen. If your data is not of this kind, there is no point in drawing up a binomial or Poisson distribution with which to compare it. Assuming, however, that an appropriate distribution has been selected and that calculations have been made of the mean and standard deviation to be ascribed to it, the frequency distribution ϕ from one of equations (13.10)–(13.12) is multiplied into the number of the sample, n, for each appropriate value of the variate. Note that although equations (13.10)–(13.12) provide formulae for calculating the frequency distribution of any value of the variate, it is usually far simpler to refer to a set of standard tables. Statistical tables are found nearly as commonly as logarithm tables; several short collections of such tables have been published and are listed in section 14.12; the tables are also found at the back of most elementary textbooks of statistics.

Example 13(g) A cytology clinic deals with 100 cervical smears each week. Over one year the number of positive (cancerous) cases per week was as given in Table 13.5, columns (1) and (2). Find the expected values to match these records if the distribution is (a) binomial, (b) Poisson, and (c) Gaussian.

Table 13.5 Incidence of positive cervical smears

Original data			Binomial		Poisson		Gaussian	
(1)	(2)	(3)	(4)	(5)	(6)	(7)	(8)	(9)
Weeks	Cases							
n_i	x_i	$n_i x_i$	$f_b(x)$	$w_b(x)$	$f_P(x)$	$w_P(x)$	$f_G(x)$	$w_G(x)$
3	0	0	0.0476	2.48	0.0498	2.59	0.0591	3.16
6	1	6	0.1471	7.65	0.1494	7.77	0.1205	6.46
12	2	24	0.2252	11.72	0.2240	11.65	0.1846	9.90
10	3	30	0.2275	11.83	0.2240	11.65	0.2129	11.41
14	4	56	0.1706	8.87	0.1680	8.74	0.1846	9.90
4	5	20	0.1013	5.28	0.1008	5.24	0.1205	6.46
2	6	12	0.0496	2.58	0.0504	2.62	0.0594	3.16
0	7	0	0.0206	1.07	0.0216	1.12	0.0218	1.16
1	8	8	0.0074	0.38	0.0081	0.42	0.0061	0.32
0	9	0	0.0023	0.012	0.0027	0.14	0.0013	0.07
52	36	156						

Answer: Detailed numerical results are listed in Table 13.5.

$$\sum_i x_i^2 = 204.$$

Mean,
$$m = \left(\sum n_i x_i\right) \Big/ n = 156/52 = 3.0.$$

S.d,
$$s = \sqrt{\frac{n\sum x^2 - \left(\sum x\right)^2}{n(n-1)}} = \sqrt{3.511} = 1.874.$$

m and s are the best estimates available for μ and σ in the ideal theoretical population distributions and so are used in the calculations.

(a) *Binomial distribution.* If a recorded positive case is considered a 'success', the expected number of successes per week is the mean, m, so that $p = 3/100$ and $q = 97/100$, since 100 cases are examined each week. For this distribution, the observed mean is used as $\mu = m = np = 3.0$, but not the observed standard deviation since the distribution requires that $\sigma = \sqrt{npq} = 1.7059$. The frequency of x successes is given by $f_b(x) = {}^{100}C_x p^x q^{100-x}$. The calculated values are listed in column (4) of Table 13.5. The expected number of weeks in the year in which x successes are recorded is then the fraction $f_b(x)$ of the total, $n = 52$, so that $w_b = 52 f_b(x)$ and these values are listed in column (5) of the table. The values of $w_b(x)$, inevitably non-integer, are to be compared with the necessarily integer values of column (1).

(b) *Poisson distribution.* The assumption of a Poisson distribution requires that the population mean, μ, is set to $m = 3.0$ in equation (13.11). This is then also the value of the population variance, so that the population standard deviation, σ, has the value $\sqrt{3} = 1.7321$. The individual frequencies are given by $f_P(x) = (e^{-3.0} \cdot 3.0^x)/x!$ and values of this quantity are shown in column (6) of Table 13.5 with the expected number of weeks in which the value x would be recorded shown in column (7). Because $n = 100$, a high value, the Poisson distribution values of columns (6) and (7) are close to those of the binomial distribution given in columns (4) and (5).

(c) *Gaussian distribution.* The Gaussian distribution has the advantage over the binomial and Poisson distributions that it allows both the mean and the standard deviation to be fitted whereas the other two fit only the mean. However, in the present case the Gaussian is an unlikely distribution since it is symmetric about the mean. The mean is to be at $x = 3$ and the distribution must extend to give significant scores at values of 7 or 8 for x, in which case it would also extend to x-values of -4 or

−5. But negative values of x are meaningless in the context of the given situation. The distribution has nevertheless been calculated just for positive values of x to illustrate the possibilities; calculated frequencies, $f_G(x)$, are given in column (8). The expected number of weeks in which x occurrences would be recorded is given in column (9) of Table 13.5 after multiplication by a factor to bring the total for the year to 52 weeks.

Summary. Columns (5), (7) and (9) of Table 13.5 show, when compared with column (1), that none of the three distributions gives a particularly good fit to the recorded data. The Gaussian distribution is somewhat too flat and too broad. The other two are better and similar to each other with the binomial distribution seemingly slightly better than the Poisson; but a quantitative test of this difference is a matter for hypothesis testing to be dealt with in Chapter 14.

EXERCISE 13D

Table 13.6 gives details of the size of pig litters in a group of farms in one year. It also gives the total number of females in each group of litters. For the distribution of litter size, discuss whether an appropriate description could be (a) a binomial, (b) a Poisson or (c) a Gaussian (normal); and if it could be so described, calculate the distribution. Might any of these distributions apply to the data on numbers of female piglets?

Table 13.6 Litter size and total number of female piglets within litters at Tealby Top Pig Farms, 1990

Number in litter	Number of litters	Total number of females
1	0	0
2	1	1
3	3	6
4	6	15
5	11	30
6	19	68
7	9	37
8	7	33
9	3	15
10	1	5
11	0	0

14

Statistics: correlation, regression and hypothesis testing

In a typical scientific experiment, you observe a number of instances in which you or an outside agency has changed one of the variables, which we can call x. As before, we have the idea that x is the independent variable – and this idea carries with it a possible implication that x is the fundamental variable, causing other things to happen. In the experiment, having chosen or set x, you then observe y. If you call y the dependent variable, you mean that whatever it measures depends on and is caused by x. Before you give this weight of meaning to y, pause to consider in a formal way what your scientific procedure is and what you should do to check it. There follows here one possible route: it is the route of this chapter and the route of some laboratory studies, but certainly there are other routes, too, to the successful completion of a research project.

1. Check that y does indeed depend on x, or possibly that they have a common cause; this is the aim of **correlation analysis**.
2. Fit the data to find the actual mathematical expression of that dependence; this is the aim of **regression analysis**.
3. With the mathematical expression provided either by regression analysis or by some independent theorizing you then have a hypothesis which gives an explanation of what has been observed, except that the hypothesis leads to predicted observations y_i, whereas what was actually observed was y'_i. The calculation of whether what you actually observed is near enough to what was expected is the aim of **hypothesis testing**.

Each of these three actions can involve elaborate calculations if all the ramifications are explored. Here we just skim through the most frequently encountered circumstances.

14.1 CORRELATION

If a variate, x, has some distribution, $\phi(x)$, the **expectation** or expected value of x is its mean value. The expectation of a quantity is usually written as the function E, so

$$E(x)=\int_{-\infty}^{\infty} x\phi(x)\,\mathrm{d}x \quad \text{or} \quad \sum x\phi(x) \tag{14.1}$$

depending on whether the distribution of x is continuous or discrete. Expectation is not yet another name for mean or average of the variable, because it can be applied to any function of x. Thus the expectation of $f(x)$ is $E[f(x)]=\int_{-\infty}^{\infty} f(x)\phi(x)\,\mathrm{d}x$.

If there are two quantities with variates x and y and if they are totally independent of each other

$$E(xy)=E(x)E(y). \tag{14.2}$$

That is, if repeated measurements are made of a pair of quantities, each with a probability distribution, and each quite independent of the other, the average value of the product of the pair equals the product of their average values. That is what common sense might lead one to expect and doesn't need further discussion.

If the two variates ξ and η in a population are not entirely independent they are said to be correlated. The extent of their correlation is measured by the extent to which the two sides of equation (14.2) are unequal. The measure is called the correlation coefficient, given the symbol ρ and defined by

$$\rho=\frac{E(\xi\eta)-E(\xi)E(\eta)}{\sigma_\xi\sigma_\eta}. \tag{14.3}$$

The denominator in equation (14.3), the product of the standard deviations of ξ and η separately, serves to standardize the correlation coefficient. As a result, the maximum value of ρ is $+1$ (perfect positive correlation of ξ with η), the minimum value is -1 (perfect negative correlation, so that ξ is perfectly correlated with $-\eta$), and between the two extremes, if ξ and η are uncorrelated and quite independent of each other, $\rho=0$ since equation (14.2) is then true. However, it must be emphasized that the correlation coefficient is a measure only of linear correlation or dependency; it can be very misleading if applied to two variates which are closely but non-linearly related.

The formula for expectation is given by equation (14.1) and the formula for the standard deviation has been given in section 13.4. Thus it is no great problem to express ρ in a more useful form than equation (14.3). Since equation (14.3) refers to the whole population whereas in the laboratory and elsewhere one usually deals with samples, we can use r for the sample correlation coefficient and the working formula is

$$r=\frac{n\sum xy-\sum x\sum y}{\sqrt{\left(n\sum x^2-\left(\sum x\right)^2\right)\left(n\sum y^2-\left(\sum y\right)^2\right)}}. \tag{14.4}$$

The correlation coefficient in the form of equation (14.4), which follows fairly easily from equations (14.3), (14.1) and (13.9), is convenient for use with an electronic calculator or computer: it can be accumulated as you go along, without having to wait until all the results have been gathered so that the means can be worked out.

Indeed – and note this advice – you shouldn't even try to work out the correlation coefficient these days unless you have a calculator or computer software package appropriate to the purpose. So examples in this case give no insight into how the answer is reached. Nevertheless, here are two cases as demonstrations of what can be done.

Example 14(a) Find the correlation coefficient for rating and number of light fittings detailed in Table 13.1.

Answer: Entering the number pairs (20, 3), (40, 10), etc. into equation (14.4) yields the value $r=-0.11$. Note that this result shows that the two variates are effectively independent.

Example 14(b) Find the correlation coefficient for stellar magnitude and (a) number of stars, and (b) logarithm of number of stars, as given in Table 13.2.

Answer:

(a) Entering the number pairs (0, 3), (1, 9), etc., into equation (14.4) yields the value $r=0.65$.
(b) Entering the number pairs (0, 0.40), (1, 0.95), etc., into equation (14.4) yields the value $r=0.99$.

Note that the low value of r in (a) comes about because the relationship between number and magnitude of stars is far from linear even though it is very strong; the strength of the relationship is shown by the graph of $\log n$ versus magnitude in Fig. 13.2 and the nearly linear relationship in this case leads to the very high value of r in (b). The contrast of values of r in (a) and (b) shows the weakness and strength of the correlation coefficient as a statistical measure.

EXERCISE 14 A

Use a calculator or computer to check the results of examples 14(a) and (b), using built-in programs or else the formula of equation (14.4).

14.2 REGRESSION: THE METHOD OF LEAST SQUARES

More often than not, experimental science produces results which don't need the evidence of a correlation coefficient. This is because the variate observed as the result of the experiment is firmly believed to depend on the value of the independent variate set in the experimental conditions. Indeed, my own view is that, in the natural sciences, a dependency that isn't self-evident from the experimental results, and that needs a correlation coefficient to reveal it, isn't worth bothering with. The first test of a belief in such a dependency is usually the plotting of a graph of y against x. The graph shows whether or not y does indeed change in a regular way as x changes and it also shows what kind of relationship there may be between x and y – linear, exponential, parabolic, etc.

However, even when theory and experiment are in the highest accord in respect of the relationship between the dependent and independent variates, there invariably remains the requirement of fitting the experimental results as well as possible to the theoretical or empirical model. That is to say, an attempt must be made to fit the data to a function $y = f(x)$. This exercise is the procedure of regression analysis.

It is generally assumed that the value of x is known exactly, without error. Not infrequently, this assumption is patently untrue; nevertheless, it continues to be made as a way of avoiding some quite difficult mathematics and tedious statistical analysis which would be required if a probability distribution of the independent variable were admitted. Take note, therefore, that the methods to be explored here are not strictly appropriate to all situations. Each actual experimental situation needs to be critically assessed for the best route by which to extract the information contained in the results. That warning having been given, I can return to the case where x is supposed to be known exactly and all the experimental or observational error is in y.

Suppose that either theory or else a graph of results leads to the supposition that the connection between x and y can be written as some known function f such that $f(x) = y$. But that is the ideal, error-free situation. It means that, if x is set to a value of exactly x_1, then y should appear with the exact value y_1, in which case

$$f(x_1) - y_1 = 0. \tag{14.5}$$

However, the real-world experiment gives a result y'_1, where $y'_1 = y_1 + \varepsilon_1$ and ε_1 is called the error. The idea of this error is slippery. If a meter in the experimental apparatus reads 12.5 and you write 13.5 in your notebook, there is a real error committed. However, if the meter reads 12.5 and is correctly recorded by you in circumstances such that you expected 13.5, there may have been a malfunction (error) in the instrumentation; but alternatively, your expectation may have been badly founded and instead of an experimental error there may be an inadequate theory or the influence of some factor hitherto overlooked. In either case, though, equation (14.5) no longer holds and instead

$$f(x_1) - y_1 = \varepsilon_1. \tag{14.6}$$

Repetitions of the experiment with the same or other values of x give other errors – $\varepsilon_2, \varepsilon_3$, etc. and the errors complicate the determination of $f(x)$. If, for example, there is reason to believe that $y = a\cos x + b\sin x + c$, the task is to find the best values of a, b, and c. In general, the task is, knowing the form of f, to find the best values of the constants in it.

This last sentence contains three traps. 'Knowing the form of' the function f really means 'believing the form of the function to be such-and-such'; and part of the purpose of the analysis is to reveal the consequences of that belief – consistency or contradiction with previous results, for example. 'Best value' means that the experimentally determined values y_i' should agree with the predicted values $f(x_i)$ within limits calculated according to some objective criteria, even though these criteria will have been subjectively selected. And finally, 'the constants' within the function f turn out to be what in other circumstances are variables, with values settled as the solutions to simultaneous equations.

The route by which this matter is nowadays invariably disposed of goes by the name of the method of least squares. It is a procedure carried out according to the following rule.

Choose the constants of the function f such that, for all the experimental results fitted into equations like equation (14.6),

$$\sum_{i=1}^{n} \varepsilon_i^2 \quad \text{is a minimum.} \tag{14.7}$$

Note that other rules are imaginable. One possibility is to take the criterion that $\sum \varepsilon_i = 0$; but this has the disadvantage that it could be made zero by one large negative value, ε_j, with all other values of ε being positive and small. This could give a function f not in accord with intuition and common sense of what is the 'best' relation, as shown by the drawn line in Fig. 14.1 which clearly ought to pass through the group of points lying uniformly above it. Or the rule could have involved $\sum \varepsilon_i^3$ or $\sum \varepsilon_i^4$ or infinitely many other functions of the set of numbers $\{\varepsilon_i\}$.

In fact, the chosen rule, the criterion of equation (14.7), is almost always adopted, but it should be understood that the choice of it is empirical. The name of the procedure, method of least squares, comes obviously from equation (14.7). Also, it has the virtue that, if the errors are true errors, fall randomly, and are normally distributed, then the 'best values' of the constants in f, as calculated by the method of least squares, are the most

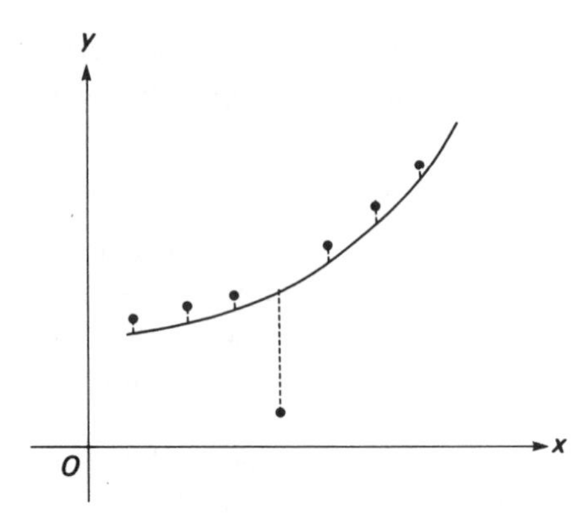

Figure 14.1

probable values. Furthermore, the method can be shown to give what is called an unbiased combination of the experimental observations, with minimum variance of the supposed errors. So, with the hesitant supposition that the method is acceptable, we can look at how it works.

14.3 LINEAR REGRESSION

The function f which is to connect x and y can be any sort of function, depending on experimental circumstances and the indications of theory; but the simplest case to understand is that in which the connection between x and y is linear (cf. Section 6.1). In these circumstances, what is expected when the independent variable is set at x_1, is $y_1 = mx_1 + c$ or, which is the same thing,

$$mx_1 - y_1 + c = 0. \tag{14.8}$$

Equation (14.8) is the linear form of equation (14.5), and, correspondingly, what is actually found, because of experimental error, is

$$mx_1 - y_1 + c = \varepsilon_1 \tag{14.9}$$

where equation (14.9) is the linear form of equation (14.6). Further experiments give the number triplets $(x_2, y_2, \varepsilon_2)$, $(x_3, y_3, \varepsilon_3)$, etc. The procedure now is to choose values for the constants m and c so that $\sum \varepsilon_i^2$ takes a minimum value. From equation (14.9)

$$\varepsilon_i^2 = (mx_i - y_i + c)^2.$$

Hence, writing S for $\sum \varepsilon_i^2$, the quantity to be minimized is

$$S = \sum_{i=1}^{n} (mx_i - y_i + c)^2. \tag{14.10}$$

In Chapter 10 the minimum value of a variable was found to be at the position where the first derivative of the variable was zero and the second was positive. In the present case, the variable S depends on two others, m and c. As a consequence, it can be shown that the minimum will occur when the partial derivatives (cf. Section 10.7) of S with respect to m and c are both zero. That is, the required conditions are that

$$\frac{\partial S}{\partial m}=0 \quad \text{and} \quad \frac{\partial S}{\partial c}=0.$$

Now, differentiating equation (14.10) with respect to m,

$$\begin{aligned}\frac{\partial S}{\partial m}&=\sum 2x_i(mx_i-y_i+c)\\ &=2\sum(mx_i^2-x_iy_i+cx_i)\\ &=2\left(m\sum x_i^2-\sum x_iy_i+c\sum x_i\right).\end{aligned}$$

And the condition $\partial S/\partial m=0$ leads to

$$m\sum x_i^2+c\sum x_i-\sum x_iy_i=0. \tag{14.11}$$

Similarly, the condition $\partial S/\partial c=0$ leads to

$$m\sum x_i+cn-\sum y_i=0. \tag{14.12}$$

Equations (14.11) and (14.12) provide a pair of simultaneous equations in m and c; such a set of equations, a pair in the present case, or more in cases of non-linear regression, which is discussed below, are called normal equations. Solving the equations as in Chapter 9 gives

$$m=\frac{n\sum x_iy_i-\sum x_i\sum y_i}{n\sum x_i^2-\left(\sum x_i\right)^2} \tag{14.13}$$

$$c=\frac{\sum x_i^2\sum y_i-\sum x_i\sum x_iy_i}{n\sum x_i^2-\left(\sum x_i\right)^2}. \tag{14.14}$$

Values of m and c as calculated by using equations (14.13) and (14.14) are the gradient (or slope) and intercept of the best-fit straight line connecting x and y. These are nearly always preferable to the values obtained from a line drawn 'by eye' on a plot of y against x, since the latter method is usually excessively influenced by the extreme points, at maximum and minimum values of x and y. Many electronic calculators offer a linear regression function, using the formulae of equations (14.13) and (14.14) so it is now fairly easy to get the calculated result.

Example 14(c) Assuming a linear relationship, find the regression equation relating weight and age of European boys between the ages of one and ten years from the data of Table 14.1.

Answer: Using an electronic calculator and the formulae of equations (14.13) and (14.14), if (weight)$=m$(age)$+c$ then $m=2.51\ \mathrm{kg\ yr^{-1}}$ and $c=7.12$ kg.

EXERCISES 14B

1. Use an electronic calculator to check the results given in example 14(c).
2. Assuming a linear relation between age and height between the ages of 3 and 15 years inclusive, according to the data of Table 14.1 for European boys, use an electronic calculator to find the regression line between age and height and so show that the gradient of the line implies an average annual increase in the boys' height in that age range of 5.80 cm.

Table 14.1 Average weight and height of young European males

Age (yrs)	Weight (kg)	Height (cm)
1	10.0	75.3
2	12.6	87.6
3	14.7	96.3
4	16.6	103.5
5	19.1	110.2
6	21.9	117.7
7	24.5	124.3
8	27.3	130.2
9	29.9	135.7
10	32.6	140.5
11	35.2	144.5
12	38.3	149.9
13	42.2	155.3
14	48.8	163.0
15	54.5	168.1
16	58.8	171.9
17	61.8	174.0
18	63.1	174.8

14.4 NON-LINEAR REGRESSION

The relationship between the independent variable x and the quantity y, subject to experimental error, is not necessarily linear; and there is, of course, a vast number of plausible non-linear relationships. In the case of a polynomial relationship, the method is an extension of that used to find the constants of a linear regression. For example, suppose that a cubic relationship is to be fitted, where

$$y=a+bx+cx^2+dx^3.$$

Then, as before, setting $x=x_1$ produces experimentally, instead of the 'true' result y_1, an erroneous result $y_1+\varepsilon_1$. Thus

$$a+bx_1+cx_1^2+dx_1^3-y_1=\varepsilon_1. \tag{14.15}$$

Squaring and summing all the equations which are like equation (14.15) gives

$$S=\sum \varepsilon_i^2=\sum (a+bx_i+cx_i^2+dx_i^3-y_i)^2.$$

The minimization of S is achieved by making

$$\frac{\partial S}{\partial a}=\frac{\partial S}{\partial b}=\frac{\partial S}{\partial c}=\frac{\partial S}{\partial d}=0.$$

This gives the normal equations, a set of four simultaneous equations, by which the four constants a, b, c, and d, may be evaluated as in Section 9.3. For example, $\partial S/\partial a=0$ leads to

$$2\sum (a+bx_i+cx_i^2+dx_i^3-y_i)=0.$$

Hence,

$$na+b\sum x_i+c\sum x_i^2+d\sum x_i^3-\sum y_i=0; \tag{14.16}$$

and $\partial S/\partial b=0$ leads to

$$2\sum x_i(a+bx_i+cx_i^2+dx_i^3-y_i)=0$$

so that

$$a\sum x_i+b\sum x_i^2+c\sum x_i^3+d\sum x_i^4-\sum x_i y_i=0.$$

Similar derivations come from $\partial S/\partial c=0$ and $\partial S/\partial d=0$. One or other of the methods discussed in Chapter 9 can then be used to solve the set of equations. And clearly the same method could be applied to regression curves of higher or lower degree.

In some cases it is possible to linearize non-linear functions. For example, if the expected regression is of the form $y=a\cdot e^{bx}$, one can write

$$\ln y=\ln a+bx.$$

Now take a new dependent variate $y' = \ln y$, and a constant $a' = \ln a$. Thus the relationship now to be fitted is

$$y' = a' + bx.$$

This is dealt with as if it were ordinary linear regression, but after taking logarithms to base e of each experimental result.

Finally, and for completeness, it is worth noting that the same method can be used if the experimental result depends linearly on a number of independent variables. Suppose, for example, that the experimental result Q depends on the (exactly known) variables x, y, and z, according to the relationship

$$Q = a + bx + cy + dz.$$

Then some particular experimental setting (x_1, y_1, z_1) yields the erroneous result $Q_1 + \varepsilon_1$, where

$$a + bx_1 + cy_1 + dz_1 - Q_1 = \varepsilon_1.$$

In this case,

$$S = \sum_i \varepsilon_i^2 = \sum_i (a + bx_i + cy_i + dz_i - Q_i)^2.$$

Once again, the constants a, b, c, and d are found by obtaining the normal equations which result from requiring that

$$\frac{\partial S}{\partial a} = \frac{\partial S}{\partial b} = \frac{\partial S}{\partial c} = \frac{\partial S}{\partial d} = 0.$$

As with the case of polynomial regression, this method of multivariate regression can be applied to situations with more or fewer independent variables.

Example 14(d) From the data of Table 13.2 for the number of stars, n_i, of magnitude m_i, put $N_i = \log n_i$ and $M_i = \log m_i$, and with the m-values 1, 2, 5, 10, and 20 find the best-fit quadratic relationship between M and N of the form $N = a + bM + cM^2$.

Answer: The data with which to work is shown in Table 14.2. The set of normal equations for the problem is

$$na + \left(\sum M_i\right)b + \left(\sum M_i^2\right)c = \sum N_i$$

$$\left(\sum M_i\right)a + \left(\sum M_i^2\right)b + \left(\sum M_i^3\right)c = \sum M_i N_i$$

$$\left(\sum M_i^2\right)a + \left(\sum M_i^3\right)b + \left(\sum M_i^4\right)c = \sum M_i^2 N_i$$

Table 14.2

m_i	M_i	n_i	N_i
1	0	9	0.95
2	0.301	29	1.46
5	0.699	1.09×10^3	3.09
10	1.000	2.07×10^5	5.32
20	1.301	4.40×10^8	8.64

where n is the number of sets of data. Using the matrix method of solution of simultaneous equations described in Section 9.5, the set of equations can be written as

$$\mathbf{T}\begin{pmatrix} a \\ b \\ c \end{pmatrix} = \mathbf{S}, \quad \text{where} \quad \mathbf{T} = \begin{pmatrix} n & \sum M_i & \sum M_i^2 \\ \sum M_i & \sum M_i^2 & \sum M_i^3 \\ \sum M_i^2 & \sum M_i^3 & \sum M_i^4 \end{pmatrix} \quad \text{and} \quad \mathbf{S} = \begin{pmatrix} \sum N_i \\ \sum M_i N_i \\ \sum M_i^2 N_i \end{pmatrix}.$$

With $n = 5$, the following values are found with an electronic calculator:

$$\sum M_i = 3.301 \qquad \sum N_i = 19.46$$
$$\sum M_i^2 = 3.272 \qquad \sum M_i N_i = 19.16$$
$$\sum M_i^3 = 3.571 \qquad \sum M_i^2 N_i = 21.59$$
$$\sum M_i^4 = 4.112$$

Hence

$$\mathbf{T} = \begin{pmatrix} 5 & 3.301 & 3.272 \\ 3.301 & 3.272 & 3.571 \\ 3.272 & 3.571 & 4.112 \end{pmatrix} \quad \text{and} \quad \mathbf{S} = \begin{pmatrix} 19.46 \\ 19.16 \\ 21.59 \end{pmatrix}$$

It is now necessary to form the transpose matrix of **T**, then the adjoint matrix of **T**, and then to divide through by the determinant formed of the elements of **T** to obtain finally the reciprocal of **T** as in Section 9.5. Writing **R** for the reciprocal, the calculations give

$$\mathbf{R} = \begin{pmatrix} 0.863 & -2.324 & 1.332 \\ -2.324 & 12.120 & -8.676 \\ 1.332 & -8.676 & 6.718 \end{pmatrix}$$

Thus, according to the account given in Chapter 9,

$$\begin{pmatrix} a \\ b \\ c \end{pmatrix} = \mathbf{R}\cdot\mathbf{S} = \begin{pmatrix} 1.010 \\ -0.296 \\ 4.704 \end{pmatrix}$$

Hence the required regression curve is

$$N = 1.010 - 0.296M + 4.704M^2.$$

EXERCISES 14C

1. Use an electronic calculator to confirm the results given in example 14(d).
2. Use an electronic calculator to show that the linear regression analysis of the data used in example 14(d) gives $N = 5.778M + 0.077$. Hence show that $S_L = 3.321$, where S_L is the sum of the squares of the differences (errors) between the predicted and observed values of N from the linear regression equation. Calculate S_Q, the corresponding quantity using the predictions of the quadratic regression equation of example 14(d) and show that its value is less than 1% of the value of S_L.
3. Use an electronic calculator and only the data of Table 14.1 for ages 3, 6, 9, 12, 15 and 18 years to find the quadratic regression equation of height (H) as a function of age (A). Calculate the predicted values of height at ages 5, 10, 15, 20, 25, 30 and 35 years and comment on the results.

14.5 HYPOTHESIS TESTING

'Hypothesis testing' is a rather grand name given by statisticians to a very frequently occurring activity in the experimental sciences, namely, answering the question 'Is *this* result compatible with *that*?' 'That' may be a group of experimental observations, or a deduction from a theory, or an expectation dredged up from a hunch. Whatever the precise form of 'that', it is referred to as a hypothesis which, in turn, is based on a description, or model, of the situation being studied. The model may be a fairly realistic account of things, or an abstract mathematical formulation, or something in between. In the statistical procedure, the hypothesis being tested (i.e. being examined for plausibility) is usually denoted by H_0.

If H_0 is tested, it is tested *against* another hypothesis, H_1, or against several hypotheses, $H_1, H_2, H_3, \ldots$ The hypothesis H_1 may provide:

1. the statement that H_0 is false; or
2. a model like that from which H_0 is derived, but with different parameters; or

3. a completely different model; or
4. some variant of these possibilities.

All data with errors – which is to say, all data in the real world – allow some probability to both H_0 and H_1. The objective of hypothesis testing is to distinguish between H_0 and H_1 by gauging the significance, with respect to them, of the data to hand. The significance is obtained in the form of the probability of having observed such a collection of data if either H_0 or H_1 is true.

14.6 THE NULL HYPOTHESIS

Suppose, for example, that the data are a set of results y_1, y_2, ... which we can write as $\{y_i\}$. Suppose also that a theory or model M_0 is being examined. Then H_0 could be the hypothesis asserting the compatibility of $\{y_i\}$ with M_0; and H_1 could assert the compatibility of $\{y_i\}$ with any other theory or model. Usually, however, there is only the one model M_0 which is of interest in the first instance. For reasons of logic and efficiency which, however, I am not going to tease out here, the most frequently adopted hypotheses are

1. H_0: the set $\{y_i\}$ is not inconsistent with M_0;
2. H_1: the set $\{y_i\}$ is inconsistent with M_0.

Such a form of H_0 is called the null hypothesis.

Testing by way of a null hypothesis is so common a practice that the hypothesis itself is usually not stated. As a consequence, the meaning of the result of the test is often obscured or misunderstood. The idea of results being 'not inconsistent' is typically taken to mean that, if the theory or model M_0 is sound, there is at least a 5% chance (i.e. a one out of twenty chance) of the set $\{y_i\}$ having been observed, and if that is the case, the hypothesis H_0 may be upheld. Alternatively, the hypothesis H_1 is typically upheld when analysis shows that, if M_0 is sound, the chance of $\{y_i\}$ having been observed is less than 5%. The calculation of the chance is the first requirement in making a test of significance, i.e. of the reliability of the hypothesis. The purpose of hypothesis testing is to give objective grounds for deciding whether to uphold H_0 or H_1.

14.7 ERRORS WHICH RESULT FROM HYPOTHESIS TESTING

As the result of a test of significance between the null hypothesis H_0 and the hypothesis H_1, four situations can arise when a conclusion is drawn, with two types of error as shown in Table 14.3. The correct decision is made if it is concluded that H_0 is true when it is actually true (and H_1 is false) or that

Table 14.3 Errors in accepting or rejecting the null hypothesis

Decision taken or conclusion drawn	Actual state of affairs	
	H_0 true	H_1 true
H_0 true	correct	Type II error
H_1 true	Type I error	correct

H_0 is false when it is actually false (and H_1 is true). But if the conclusion is that H_0 is false (and H_1 true) when actually H_0 is true, a Type I error is committed. And if the conclusion is that H_0 is true (and H_1 false) when actually H_0 is false, a Type II error is committed. Thus an error of Type I rejects a true hypothesis and an error of Type II affirms a false hypothesis.

Suppose, for example, that a woman has a cervical smear test and the cells from her cervix are examined to test the hypothesis 'H_0: the cells have abnormalities not inconsistent with her having cervical cancer'. The health authorities want the correct decision: either that the woman doesn't have cancer when she truly doesn't (about 98% of cases), or that she does have cancer if she indeed does have it (about 2%). The Type I error is to conclude that she doesn't have cancer when really she does; this is sometimes called a false negative conclusion. The Type II error is to conclude that she does have cancer when really she doesn't; which is sometimes called a false positive conclusion.

In the case of the woman examined with respect to the hypothesis that she has cancer, a Type I error (which would allow the cancer to grow without treatment) is clearly much more dangerous than a Type II error (which would normally lead to a second test for confirmation). However, for a man being tried in court on a charge (hypothesis) of having committed a serious crime, a Type II error – that he is guilty and should be punished, when actually he is innocent – is much more dangerous than a Type I error – that he is innocent and should go free, when actually he is guilty.

In some situations, the avoidance of one or other type of error is as important as making a correct decision. Methods have been devised for taking such requirements into account, but even without these constraints, very many tests are available for assigning significance to hypotheses. Furthermore, the application of these tests involves the use of figures culled from statistical tables. To avoid over-elaboration, I shall suppose that you have access to statistical tables of this kind and shall deal only very briefly with three of the most commonly used tests. (A list of some books of statistical tables is given at the end of this chapter.)

14.8 DEGREES OF FREEDOM

Each of the tests to be described has an outcome depending on what is referred to as the number of degrees of freedom in the data used to test the hypotheses H_0 and H_1. The concept of number of degrees of freedom is tricky. Every independent experiment is, in a manner of speaking, a free and unfettered attempt at determining the values of the parameters of the model M_0 being examined. In the first instance, if n independent experiments are carried out there are n such free attempts, or n degrees of freedom, in the determination.

However, if the procedure of hypothesis testing involves, either directly or indirectly, the mean of the n results, then the number of degrees of freedom is reduced by one. This is because, if you were given $n - 1$ results and the value of the mean, you could calculate the value of the nth result; it would no longer be independent of the data to hand. The mean is indirectly involved if the analysis uses the standard deviation or variance of a sample since these are calculated with reference to the mean.

In some cases the number of degrees of freedom is further reduced by constraints imposed in the presentation of the data. For example, consider a trial in which cattle are given different dietary or veterinary treatments, either separately or together, and the results recorded are the numbers of animals giving low, medium, and high milk or meat yields. If a contingency table is prepared with a different treatment in each of r rows and a different outcome in each of c columns, the various entries must be reconciled so that the total of the numbers in the rows equals the total of the numbers in the columns. Hypothesis testing usually consists of comparing the table of observed results with a table prepared according to some model. Typically (but not always) the number of degrees of freedom remaining in these circumstances is $(r-1)(c-1)$. This value comes from consideration of the number of superfluous items of data when the used information includes the marginal subtotals of the rows and columns (cf. Table 14.4 and example 14(e) below).

Each non-standard case must, in practice, be considered in its own terms, probably after reference to a detailed account in a textbook of statistics. In the accounts which follow here, the number of degrees of freedom is indicated for the circumstances most frequently encountered.

14.9 THE χ^2 TEST

The χ^2 (chi-squared) test is typically applied in the following circumstances. An experiment or data-gathering exercise consists of n observations which fall into k mutually exclusive possible outcomes with frequencies $A_1, A_2, \ldots, A_k$; a model or theory M_0 predicts frequencies $B_1, B_2, \ldots, B_k$;

and the null hypothesis to be tested is H_0: the set $\{A_i\}$ is not inconsistent with the predictions of M_0. It is assumed, usually without thought, that any experimental errors give the observed frequencies, A_i, a normal distribution about their true values, whether or not the latter are as predicted by M_0.

The chi-squared critical value is calculated as

$$\chi_\nu^2 = \sum_i \frac{(A_i - B_i)^2}{B_i}, \qquad \text{i.e.} \qquad \sum_i \frac{(\text{observed} - \text{expected})^2}{\text{expected}} \tag{14.17}$$

where ν is the number of degrees of freedom. If the outcomes are quite independent and the only constraint is that $\sum \phi_i = 1$, where ϕ_i is the probability of the outcome which has been given the suffix i (cf. section 13.11 for the calculation of ϕ_i, and hence B_i), then $\nu = k-1$. If the outcomes are set out in a contingency table with r rows and c columns (with $rc = k$) and the subtotals of the rows and columns are reconciled, then $\nu = (r-1)(c-1)$. More degrees of freedom may be lost if the frequencies A_i are subject to further restraint.

When χ_ν^2 has been calculated according to equation (14.17), its value is compared with the value in χ^2 tables for ν degrees of freedom and at a previously chosen level of probability α. Suppose that α is chosen as 5% and that the corresponding entry in the table is $K(\nu, 5\%)$.

If $\chi^2 > K(\nu, 5\%)$, the *interpretation* is that, if the model M_0 were correct, there would be less than a 5% (i.e. less than a one in twenty) chance of having recorded a set of data like $\{A_i\}$ so divergent from the predicted set $\{B_i\}$. The *decision* then taken is usually that the hypothesis H_0 must be rejected in favour of H_1 (i.e. that the set of observed frequencies $\{A_i\}$ is indeed inconsistent with M_0); and that since the facts (i.e. the set $\{A_i\}$) are inescapable, the model M_0 must be discarded. (Except that if the experimenter is very attached to the model M_0 there is a strong temptation to repeat the experiment in the hope of getting a less discouraging set of results.) The smaller the value of α which can be taken while leaving $\chi^2 > K(\nu, \alpha)$, the more confidently can the decision to discard M_0 be taken.

If, on the other hand, $\chi^2 < K(\nu, 5\%)$, the interpretation is that, if M_0 were correct, there would have been at least a 5% chance of finding a set of data $\{A_i\}$ differing by the observed amounts from the predicted values $\{B_i\}$. The decision usually then taken is first to accept H_0 and secondly to accept M_0 as an accurate model of the situation. The larger the value of α which can be taken while leaving $\chi^2 < K(\nu, \alpha)$, the more confidently can the decision be taken.

The χ^2 tables can also be applied to the formation of confidence intervals for the standard deviation of a set of observations. Reference should be made to a statistics textbook for the details and methods of this application.

Example 14(e) Patients with a particular condition were divided into two groups and were given either treatment P or treatment Q. The patients were all initially in about the same state of sickness and the outcomes were as given in Table 14.4. Use the χ^2 test to estimate whether or not the two treatments were significantly different.

Answer: It is always most efficient to take the conservative null hypothesis 'H_0: the data are not inconsistent with there being no significant difference between the two treatments'. Thus the model being tested is that the two treatments are equally effective; but if that is the case, the best estimate of the proportion of patients who might be expected to improve is 42 (i.e. all the patients in column 2, whatever treatment was received) out of 105. Thus the number of patients expected to improve after treatment P is $(42/105)\times 60=24.0$, and the number after treatment Q is $(42/105)\times 45=18.0$. Similarly, the best estimate of the proportion not showing change is 45/105 and that of the proportion which might be expected to worsen is 18/105. Hence, the expectations in Table 14.5 can be calculated.

For each position in the table there is now an observed and an expected value. It is only among the expected values in the 'ideal' or 'theoretical' table that there are fractions of a patient in the individual entries, but the marginal totals remain the same as for the original data, which is what reduces the number of degrees of freedom. With 2 rows and 3 columns the remaining number of degrees of freedom is $(2-1)(3-1)=2$. Calculating χ^2

Table 14.4 Outcome of patient treatments; number of patients in each category

Treatment	Patient appeared to improve (I)	No apparent change in condition (N)	Patient appeared to worsen (W)	Total
P	27	21	12	60
Q	15	24	6	45
Total	42	45	18	105

Table 14.5 Hypothetical results for the treatments of equal effectiveness

	I	N	W	Total
P	24.0	25.8	10.2	60.0
Q	18.0	19.2	7.8	45.0
Total	42.0	45.0	18.0	105.0

from the elements of the first row, and then of the second, results in

$$\chi_2^2 = \frac{(27-24)^2}{27} + \frac{(21-25.8)^2}{21} + \frac{(12-10.2)^2}{12} + \frac{(15-18)^2}{15} + \frac{(24-19.2)^2}{24} + \frac{(6-7.8)^2}{6} = 3.801.$$

Tables show that the value of χ_2^2 at the 5% level of significance is 5.991. The calculated value of χ^2 in the present case is much less than this; in fact it is less than the value of χ_2^2 at the 30% level of significance. Thus there is a chance of better than one in three of obtaining results differing from those predicted by at least as much as the figures in Table 14.4. Thus the decision can be made to accept H_0 together with its implication that the treatments *P* and *Q* are equally effective.

EXERCISE 14D

A die is thrown 100 times with the results shown in Table 14.6. Can the die reasonably be considered to be unbiased? (Hint: take as the null hypothesis 'H_0: the data are not inconsistent with the die being unbiased'. On that basis, find the average expected frequency for each possible fall of the die and so calculate χ^2. There are obviously in this case 5 degrees of freedom and tables give a value of 11.070 for χ^2 at the 5% level of significance.)

Table 14.6 Throws of a die

Number	1	2	3	4	5	6
Frequency	18	11	16	17	11	27

14.10 THE *t*-TEST

The *t*-test (using what is often called Student's *t*-distribution, because its inventor wrote under the pseudonym 'Student') is used especially with values (means, etc.) obtained from samples of relatively small size (e.g. less than 25). The *t*-distribution is symmetric about its central value, which corresponds to 100% confidence or a probability of 1 (i.e. absolute certainty) in the value being studied.

There are two situations in which it is frequently used. One is to establish confidence limits of sample parameters. In this case, knowing the value of an observed quantity – a datum known with certainty – an estimate is required of the chance that the true population value of that parameter may be either above or below the observed value, so probabilities obtained from either side of the central value of the t-distribution must be used. The test is then said to be two-sided or two-tailed. The other frequent use of the t-distribution is to test the model that two samples come from the same population or from two populations with identical means. In this case, the samples provide two means, of which one is certainly larger than the other and an estimate is required of the chance of finding the observed difference between the two. Such an estimate is made from the probability obtained from just one side of the t-distribution. The test is then said to be one-sided or single-tailed. (For large-sized samples, the t-distribution becomes indistinguishable from the normal distribution.)

Confidence limits of the mean

Suppose a sample of n observations has a mean $\bar{x}$. The question may arise 'Is this result compatible with the population having a mean μ?'. This is something like the question being asked when the χ^2 test was applied, but now the contingency table is reduced to one entry and there would be no degrees of freedom for χ^2. The t-distribution makes use of the v degrees of freedom provided by the original n observations but reduced by 1 in establishing the mean $\bar{x}$, so that $v=n-1$. The value of t_v is given by

$$t_v=\frac{\bar{x}-\mu}{s/\sqrt{n}} \tag{14.18}$$

where s is the standard deviation of the sample and it is assumed that the experimental errors are normally distributed about the value $\bar{x}$. Reference to tables of the t-distribution then gives a value of the percentage confidence which can be given to the statement that $\bar{x}$ is not significantly different from μ. Alternatively, if $t_v(\alpha)$ is the value read from the table of t with v degrees of freedom and a confidence of α, then it can be said with that amount of confidence that the population mean lies within limits written as

$$\bar{x}-t_v(\alpha)s/\sqrt{n}<\mu<\bar{x}+t_v(\alpha)s/\sqrt{n}. \tag{14.19}$$

Two matters should be noted here. Firstly, the procedure really remains a test of the null hypothesis 'H_0: the experimental resuls are not inconsistent with M_0, where M_0 is the model that the sample of n is from a population of mean value μ'. Secondly, as noted above, the t-distribution tends to the Gaussian or normal distribution as n becomes large.

Identity of two population means

Suppose that a sample of size n_1 has a mean $\bar{x}_1$ and standard deviation s_1 and a second sample of size n_2 has a mean $\bar{x}_2$ and standard deviation s_2. It is often necessary to test the null hypothesis 'H_0: the data are not incompatible with the two samples having been drawn from the same population'. Alternatively, and not quite the same thing, the null hypothesis might be 'H_0: the data are not incompatible with the two samples having been drawn from two different populations which, however, have the same mean' and should have the same variance, although this requirement is often overlooked. Another possibility is to test the model that the n_1 observations with mean $\bar{x}_1$ were sampled from a population of mean μ_1, different from the population of mean μ_2 which provided the second sample of n_2 observations with mean $\bar{x}_2$. There are still other variants which might be tested, but it is unnecessary to write them all out.

The number of degrees of freedom is $\nu = n_1 + n_2 - 2$ and

$$t_\nu = \frac{(\bar{x}_1 - \bar{x}_2) - (\mu_1 - \mu_2)}{s\sqrt{(1/n_1) + (1/n_2)}} \tag{14.20}$$

where s is the standard deviation of all $n_1 + n_2$ observations and is given by

$$s = \sqrt{\frac{(n_1 - 1)s_1^2 + (n_2 - 1)s_2^2}{n_1 + n_2 - 2}}. \tag{14.21}$$

Of course, in the case where the model being tested includes the two means being equal, $\mu_1 = \mu_2$, the second bracket in the numerator of equation (14.20) is reduced to zero.

Example 14(f) A series of seven experiments on the melting of a bacterial DNA gave the melting temperature T_m as 83.5°C with a standard deviation (s.d.) of ±4.3°C. Is this result compatible with a previously published value of $T_m = 87.0$°C for DNA from the same species? A supposed repeat set of experiments with five determinations of T_m gave a result 80.7°C with s.d. ±2.1°C. Is the second set compatible with the first set?

Answer: To examine whether 83.5 ± 4.3 ($n = 7$) is compatible with a population mean of 87.0, use equation (14.18) and the null hypothesis 'H_0: the results are not inconsistent with a population mean of 87.0'. In this case $\nu = n - 1 = 6$ and

$$t_6 = \frac{83.5 - 87.0}{4.3/\sqrt{7}}$$

$$= 2.154.$$

(Only the absolute value of t is used.) Tables give 1.943 for the t-distribution with 6 degrees of freedom at the 5% level of significance in the 'one-tail' case and a value of 2.447 for the 2.5% level of significance. (The observed mean is on one side only of the supposed population mean; there is no question of asking if it is either too high or too low, so only half of all possible discrepancies are relevant and, as noted above, this leads to use only of the values given on one side, or one tail, of the t-distribution curve.)

Since the calculated value of $t_6 > 1.943$, the interpretation of the result is that there is a chance of less than 1 in 20 (although more than 1 in 40, since $t_6 < 2.4$) of obtaining a mean so far from 87.0. This is usually taken as sufficient ground for the decision to reject H_0. Thus *either* the true population mean of T_m for DNA of the bacterial species in question is not 87.0 *or* there are overlooked factors in the situation (preparative methods, species identification, etc).

To compare the two sets of experiments, use equations (14.20) and (14.21) and the null hypothesis 'H_0: the data are not incompatible with the two samples having been drawn from the same population'. Thus from equation (14.21)

$$s = \sqrt{\frac{6 \times 4.3^2 + 4 \times 2.1^2}{10}}$$

so $s = 3.586$. Also $\nu = 10$ and $\mu_1 = \mu_2$ so equation (14.20) gives

$$t_{10} = \frac{83.5 - 80.7}{3.586\sqrt{\frac{1}{7} + \frac{1}{5}}} = 1.333.$$

Again using the 'one-tail' values of the t-distribution, but now with 10 degrees of freedom, tables give values of 1.812, 1.372 and 1.093 for t at the 5%, 10% and 15% levels of significance, respectively. The calculated value, lying between the 10% and 15% significance values from the tables, shows that there was about a 1 in 9 chance of having found results with the observed discrepancy. Such an inference is usually taken as ground for accepting H_0. However, it is not strong ground for accepting H_0 with confidence, and the only way out for the conscientious scientist is to collect more experimental evidence before making a final decision.

EXERCISE 14E

Five measurements of the coagulation time of a patient's venous blood gave the result 6.3 min with a standard deviation of ± 1.4 min. A single measurement on the next day gave the value 8.0 min. Is that value compatible with the original set of results? (Hint: use equations (14.20) and (14.21) with $n_2 = 1$ and $s_2 = 0$, or else use equation (14.19).)

14.11 THE F-TEST

Suppose that two samples are drawn from normally distributed populations. The first sample of n_1 items (with $f_1 = n - 1$ degrees of freedom) has a mean

$\bar{x}_1$ and a standard deviation s_1 and the second of n_2 items (with $f_2 = n_2 - 1$) has mean $\bar{x}_2$ and standard deviation s_2. The t-distribution is used to compare the means and the probability of the samples having been drawn from populations with the same mean. But different populations may have the same mean yet differ in standard deviation, skewness, etc. To compare the standard deviations of two samples, we write

$$F = \frac{s_1^2}{s_2^2} \quad \text{or} \quad \frac{s_2^2}{s_1^2}, \quad \text{whichever is the greater.} \tag{14.22}$$

Values of the F_{f_1, f_2}-distribution are given in tables and the value of F can then be used to test the null hypothesis 'H_0: the data are not inconsistent with the model M_0, that the populations from which they are drawn have the same standard deviations'.

However, a more frequently used application of the F-test is also more interesting. If you collect two sets of data, there will inevitably be differences between the means $\bar{x}_1$ and $\bar{x}_2$ and between the standard deviations s_1 and s_2. To answer the question (put by way of the null hypothesis) of whether the two sets of results are compatible, the data are submitted to a t-test which, essentially, provides a probability for the difference of means, $\bar{x}_1 - \bar{x}_2$, after normalization (i.e. after scaling to a standard level) by use of s_1 and s_2. Suppose now that a third and fourth and still more sets of data are collected. As ever, no two means are the same, and no two standard deviations, either, but the question remains, can it reasonably be claimed that the separate samples were all drawn from the same population so that all the results can be considered mutually compatible? Although the χ^2 test can be applied to this problem, the recommended method is the application of the F-test, sometimes also referred to as an analysis of variance, which in turn is sometimes given the acronym ANOVA.

Suppose, in order to be definite, that there are four sets or groups of results, as in Table 14.7. The supposition or model, M_0, is that each group is a sample from the same population. The question then is whether the groups differ among themselves by more than might be expected in four samples. This is a question about differences *between* the groups; and it can be answered only by knowing the population's standard deviation (or the variance, which is the square of the standard deviation). However, it is no use getting the population's variance from the same differences between the means that are being questioned. To obtain an independent estimate of the population variance (or standard deviation) the differences *within* the groups are examined.

The F-test works with variances rather than with standard deviations. When the test is used to compare several sets of results, the test uses the ratio of the between-group variance to the within-group variance. Recall

Table 14.7 Fall in diastolic blood pressure (mm Hg) over a three-month period of patients at four treatment centres

Centre	*A*	*B*	*C*	*D*
	5.7	2.7	1.4	6.8
	6.8	4.7	−2.2	7.2
	2.4	8.2	4.0	4.1
	4.3	7.2	1.1	2.1
	1.8	2.4	5.0	3.8
		3.6	5.2	
		6.2	3.9	
			1.6	
n_j	5	7	8	5
$\bar{x}_j$	4.2	5.0	2.5	4.8
s_j	2.123	2.259	2.502	2.153

from section 13.6 that the variance is a sum of squares (namely, squares of difference of statistic from mean) divided by the number of degrees of freedom. Since there may not be the same number of degrees of freedom for the between-group variance as for the within-group variance, there are two values for the number of degrees of freedom for each member of the *F*-distribution family, unlike the distribution families of χ^2 and t, each of which relates only to one number of degrees of freedom.

The method of administering the test falls into a fairly obvious pattern. With x_{ij} denoting the ith element of the jth group, $\bar{x}_j$ the mean of the jth group, n_j the number of items in the jth group, and X the grand mean of all the observations, the steps are as follows.

1. Set up the null hypothesis and decide on the level of significance at which it is to be accepted or rejected (e.g. 5% or 1%).
2. Calculate the between-group variance, s_1^2, as

$$s_1^2 = \frac{\sum_j n_j(\bar{x}_j - X)^2}{f_1} \tag{14.23}$$

where f_1 is the number of degrees of freedom in this part of the calculation (so that f_1 = number of groups − 1) and

$$X = \frac{\sum_j n_j\bar{x}_j}{\sum n_j}.$$

3. Calculate the within-group variance, s_2^2, as

$$s_2^2 = \frac{\sum_j \sum_i (x_{ij} - \bar{x}_j)^2}{f_2} = \frac{\sum_j (n_j - 1) s_j^2}{f_2} \qquad (14.24)$$

   where f_2, the number of degrees of freedom in this part of the calculation, is given by $f_2 = \sum_j (n_j - 1)$.
4. Form $F_{f_1, f_2} = s_1^2/s_2^2$ or $F_{f_2, f_1} = s_2^2/s_1^2$, whichever is the larger.
5. Find the F-value from the F-distribution tables for the appropriate numbers of degrees of freedom and level of significance. (Note that many published tables provide F-values only for 5% and 1% levels of significance. Furthermore the tables usually give only F-values greater than 1, which is the reason for defining F by equation (14.22), according to whether s_1^2/s_2^2 or s_2^2/s_1^2 is used, the criterion for rejecting H_0 is that the calculated value, F_{f_1,f_2} or F_{f_2,f_1}, is greater than the tabulated value or less than it, respectively.)
6. Draw the indicated inferences and make a decision with respect to the model M_0.

It is usually most convenient to make a tabulation of the results, but the simplest way of clarifying the procedure is by way of an example.

Example 14(g) Groups of patients at four centres were put on a regime for dealing with hypertension. Attempts were made to select similar patients in each group and to apply the same regimen of treatment. The results (fall in diastolic blood pressure after three months) are recorded in Table 14.7. Do the differences between the means of the four groups justify supposing either that the groups of patients were not similar or that there were significant differences between the treatment regimes in the four centres?

Answer:

1. Set up the null hypothesis:

   H_0: the data are not inconsistent with the model of all the samples having been drawn from the same population. (NB: this wording of the null hypothesis encompasses the similarity of the patients in the four groups and also the treatment at the four centres.)

   Take, as usual, the 5% probability as the criterion of acceptability of the null hypothesis.
2. Calculate the means and standard deviations in order to be able to use equations (14.23) and (14.24). The results of these calculations are given at the bottom of Table 14.7.

3. Calculate the between-groups variance, using equation (14.23)

$$s_1^2=\frac{5(0.2)^2+7(1.0)^2+8(1.5)^2+5(0.8)^2}{4-1}=9.47.$$

In this calculation, the denominator, the first number of degrees of freedom, f_1, is one less than the number of groups.

4. Calculate the within-groups variance, using equation (14.24)

$$s_2^2=\frac{(1.5)^2+(2.6)^2+\cdots+(2.3)^3+\cdots+(1.1)^2+\cdots+(2.0)^2+\cdots}{25-4}$$

$$=5.29.$$

The denominator in the calculation of s_2^2 is the total number of statistics (patients) less the number of means determined (i.e. less the number of groups).

5. Hence $F_{3,21}=s_1^2/s_2^2=9.47/5.29=1.79$.
6. Tables of the F-distribution give the value of F with 3 and 21 degrees of freedom at the 5% level of significance as 3.072.
7. Since the calculated value of F is much less than the tabulated value, the inference is that there is an acceptable probability to H_0 being correct. In that case, it is reasonable to make the decision that the groups have, indeed, all been drawn from the same population.

The results of the calculations are summarized in Table 14.8. The conclusion that the results are compatible with each other may seem surprising when the mean of group C is only about half the means of groups A, B, and D. The reason is that the standard deviations are all so large, resulting in a large overlap between the 95% confidence limit ranges of the means.

Table 14.8 Analysis of variance: treatment of patients with hypertension

Source	Degrees of freedom (f)	Sum of squares (SSq)	Variance	Variance ratio (F)
		$\sum n_j(x_j-X)^2$	$s^2=\dfrac{SSq}{f}$	$\dfrac{s_1^2}{s_2^2}$
Between groups	$f_1=4-1=3$	28.40	$\dfrac{28.4}{3}=9.467$	$\dfrac{9.47}{5.29}=1.791$
Within groups	$f_2=4+6+7+4=21$	111.00	$111.0/21=5.286$	

EXERCISES 14F

1. Use an electronic calculator to check the calculations in example 14(g) and confirm the F-value from a table of the F-distribution.
2. Attempts are made to measure the rate constant of a dehydrogenase enzyme from a small insect in short supply. Four preparations were made of very small quantities of the enzyme with some doubts about the purity of the products. The results of the measurements are given in Table 14.9. Can the four samples be regarded as samples of equal purity of the same enzyme?

Table 14.9 Measured values of rate constants of a dehydrogenase enzyme. (Units: $s^{-1} \times 10^{-6}$)

Preparation			
1	2	3	4
2.3	2.2	2.7	2.3
2.2	2.1	2.7	2.3
2.5	2.3	2.6	2.2
2.0	2.0	2.4	2.1
2.5	2.1	2.6	2.0
	2.4		2.1
	2.3		2.4
			2.2

14.12 BIBLIOGRAPHY OF STATISTICAL TABLES

1. Kokoska, S. and Nevinson, C. (1984) *Statistical Tables and Formulae*, Springer-Verlag, Berlin.
2. Lindley, D. V. and Scott, W. T. (1984) *New Cambridge Elementary Statistical Tables*, Cambridge University Press.
3. Neave, H. R. (1981) *Elementary Statistical Tables*, Allen & Unwin, London.
4. Rohlf, F. J. and Sokal, R. R. (1981) *Statistical Tables*, 2nd edn, W. H. Freeman, New York.
5. White, J., Yeats, A. and Skipworth, G. (1979) *Tables for Statisticians*, Stanley Thorne, Cheltenham.

For I am one of the unpraised, unrewarded millions without whom Statistics would be a bankrupt science. It is we who are born, who marry, who die in constant ratio; who regularly lose so many unbrellas, post just so many unaddressed letters every year.

Logan Pearsall Smith, *Trivia*

15

Fourier series and transforms

There are many wavy or periodic occurrences in all branches of science – waves in sea and sand; bursts of sound and light; growth and metabolic cycles which are annual, monthly or daily; structures that are repetitive on the macroscopic scale (as in snakes and feathers and some altocumulus cloud formations), on the microscopic scale (as in ciliated cell surfaces) or on the atomic scale (as in crystalline structures or various macromolecular superstructures like condensed chromosomes); and many others. Quantitative analysis of these phenomena requires a mathematical technique sensitive to periodicity and mathematical functions appropriate to the purpose. These are provided by Fourier analysis. Although this may be one of the few mathematical topics in this book which you have not met before, you will find that for the most part it merely brings together other matters dealt with in earlier chapters and of itself introduces few new mathematical concepts.

The nature of the situation to be dealt with is partly illustrated by Fig. 15.1. The upper graph is of the sine function. That is, with amplitude A on

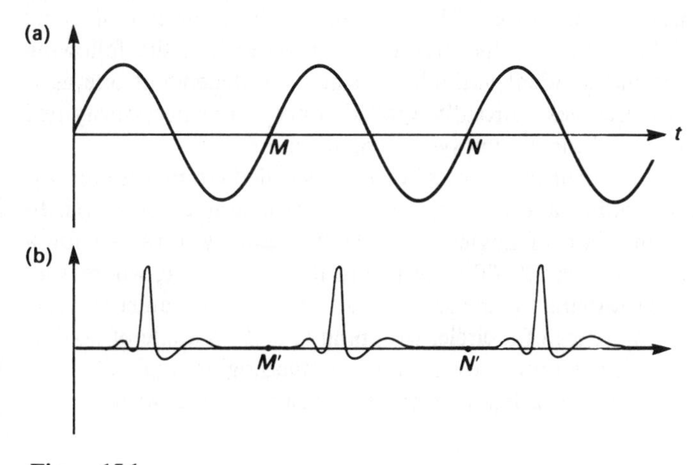

Figure 15.1

the ordinate axis and time t on the abscissa, the graph in Fig. 15.1(a) shows $A(t)=\sin 2\pi ft$ where the constant f is called the frequency. Frequency is the repetition rate of the function – the number of repeats per unit time. For periodic functions like A, f^{-1} is called the period of the function; we can write $f^{-1}=\tau$. The period of a periodic function, like the sine and other trigonometric functions (cosine, tangent, etc.) is the duration of each individual repetition because it gives the change of variable required to bring the function back to its initial value. That is, $A(t+\tau)=A(t)$, since $\tau=f^{-1}$, and $\sin 2\pi f(t+f^{-1})=\sin(2\pi ft+2\pi)=\sin 2\pi ft$.

Now look at Fig. 15.1(b) which is a trace from an electrocardiograph so that its amplitude, A', shows the change in time of the electrical signal associated with heart activity. It is clear that A' depends in a definite and even predictable way on time – the changes from moment to moment in A' are certainly not random so we could write $A'=h(t)$ where h is the heart function, quantifying the relation between A' and t. It is also clear that h is a periodic function of time, with a frequency f' such that $h(t+\tau')=h(t)$ where $\tau'=(f')^{-1}$. However, the mathematical nature of h is probably very complicated. The graph between M and N in Fig. 15.1(a) is a repeating section of the well-known shape of the sine and cosine functions; but the graph between M' and N' in Fig. 15.1(b), a corresponding repeating section of the electrocardiogram, is a shape of no known mathematical function. So whereas we know how to add, substract, multiply, differentiate, integrate, and generally manipulate the sine function of Fig. 15.1(a), we don't know the way to do so for the function $h(t)$ of Fig. 15.1(b).

The methods of Fourier analysis show how to express a given function $h(t)$ in terms of sines and cosines of different frequencies. In this instance, the independent variable, t, is a measure of time. But the techniques of Fourier analysis apply equally to situations where the independent variable represents measures in space, or temperature (one of Fourier's own first applications) or mass, or whatever. The frequencies then, instead of being temporal, would be spatial or thermal, etc. For generality, the following account leaves the independent variable as x and the dependent one as y. Frequency, f, now becomes a broadly applied concept, not necessarily tied to time or distance, and merely implying repetitiveness.

It is also often convenient to make use of what is called circular frequency, often given the symbol ω and $\omega=2\pi f$, so that, for example, $\omega t=2\pi ft$. In terms of circular measure of angles, ft must increase by 1 radian for a function like $\sin(2\pi ft)$ or $\cos(2\pi ft)$ to return to its initial value, whereas ωt must increase by 2π radians for the same result. It is as if ωt must traverse the complete circumference of a circle, wrapping round an angle of 360° at the centre, whereas ft need only cover an arc subtending an angle of about 57.3° at the centre, which is a less convenient measure to deal with.

15.1 ORTHOGONAL FUNCTIONS

If there is a sequence of functions $\psi_1(z)$, $\psi_2(z)$, $\psi_3(z)$, ..., formulated according to some rule, which do not vanish everywhere in a range of z going from a to b, the functions are said to be orthogonal in that range if

$$\int_a^b \psi_m(z)\psi_n(z)\,dz=0 \qquad \text{when } m\neq n.$$

The word 'orthogonal' means that there is a kind of independence or lack of correlation between the functions to which it is applied, but that is not a matter we need to pursue here. The only functions we need at present to be concerned with are those of the kind $\cos mx$ and $\sin mx$ for x between $-\pi$ and $+\pi$.

To show that the functions $\cos mx$ for $m=1, 2, 3, \ldots$ are orthogonal in the range $-\pi<x<\pi$, consider, when $m\neq n$,

$$\begin{aligned}
I_1 &= \int_{-\pi}^{\pi} \cos mx \cos nx\,dx \qquad (15.1)\\
&= \int_{-\pi}^{\pi} \tfrac{1}{2}[\cos(m+n)x+\cos(m-n)x]\,dx\\
&= \frac{1}{2}\left[\frac{\sin(m+n)x}{m+n}+\frac{\sin(m-n)x}{m-n}\right]_{-\pi}^{\pi}\\
&=0, \qquad \text{since } \sin k\pi=0 \text{ for integer } k.
\end{aligned}$$

Hence $\cos mx$, for integer values of m, provides an orthogonal set. We also have, for $m\neq n$, and using equation (4.8),

$$\begin{aligned}
I_2 &= \int_{-\pi}^{\pi} \cos mx \sin nx\,dx \qquad (15.2)\\
&= \int_{-\pi}^{\pi} \tfrac{1}{2}[\sin(m+n)x-\sin(m-n)x]\,dx\\
&= \frac{1}{2}\left[-\frac{\cos(m+n)x}{m+n}+\frac{\cos(m-n)x}{m-n}\right]_{-\pi}^{\pi}\\
&=0, \qquad \text{as before.}
\end{aligned}$$

Hence the set of cosine functions is also orthogonal to the set of sine functions.

In the case where m and n are equal,

$$I_3 = \int_{-\pi}^{\pi} \cos mx \cos mx \,\mathrm{d}x = \int_{-\pi}^{\pi} \cos^2 mx \,\mathrm{d}x \tag{15.3}$$

$$= \int_{-\pi}^{\pi} \tfrac{1}{2}(\cos 2mx + 1)\,\mathrm{d}x$$

$$= \frac{1}{2}\left[\frac{\sin 2mx}{4m} + x\right]_{-\pi}^{\pi}$$

$$= \pi.$$

EXERCISES 15A

1. Show that if $I_4 = \int_{-\pi}^{\pi} \sin mx \sin nx \,\mathrm{d}x$, then $I_4 = 0$ for $m \neq n$ and $I_4 = \pi$ for $m = n$.
2. Show that $I_2 = \int_{-\pi}^{\pi} \cos mx \sin nx \,\mathrm{d}x = 0$ if $m = n$.

15.2 FOURIER SERIES

Suppose that $f(x)$ is a periodic function of x with period 2π so that $f(x+2\pi) = f(x)$. Then, provided that $f(x)$ fulfils certain other conditions described below, which it usually does, Fourier's theorem states that $f(x)$ can be expanded as a series of sines and cosines:

$$f(x) = a_0 + \sum_{n=1}^{\infty} (a_n \cos nx + b_n \sin nx) \tag{15.4}$$

where the coefficients a_n and b_n are (real) constants. If equation (15.4) is multiplied through by $\cos mx$ and integrated with respect to x between the limits $-\pi$ and π, then equations (15.1) and (15.2) show that all the terms on the r.h.s. vanish except for the one containing a_m. From equation (15.3) it follows that

$$a_m = \frac{1}{\pi} \int_{-\pi}^{\pi} f(x) \cos mx \,\mathrm{d}x \tag{15.5}$$

Similarly, using the result of exercises 15A, number 1,

$$b_m = \frac{1}{\pi} \int_{-\pi}^{\pi} f(x) \sin mx \,\mathrm{d}x, \tag{15.6}$$

and integrating equation (15.4) directly, since

$$\int_{-\pi}^{\pi} \cos mx \,\mathrm{d}x = \int_{-\pi}^{\pi} \sin mx \,\mathrm{d}x = 0,$$

it follows that

$$\int_{-\pi}^{\pi} f(x)\,dx = \int_{-\pi}^{\pi} a_0\,dx = 2a_0\pi$$

so that

$$a_0 = \frac{1}{2\pi}\int_{-\pi}^{\pi} f(x)\,dx. \tag{15.7}$$

Some textbooks give $a_0/2$ in place of a_0 in equation (15.4) so the factor of 2 is lost from the denominator of equation (15.7) which defines a_0; the resulting value of a_0 then matches equations (15.5) and (15.6) which define a_n and b_n. However, the advantage of the definition given here is that a_0 is now the mean or average value of $f(x)$ over the range $-\pi$ to π. This follows since the distribution function of $f(x)$ is flat (rectangular) (cf. Section 13.10) and equal to $(2\pi)^{-1}$. Writing equation (15.7) in the form

$$a_0 = \int_{-\pi}^{\pi} \frac{1}{2\pi} f(x)\,dx$$

the identity of a_0 with the mean (or expectation) of $f(x)$ is apparent.

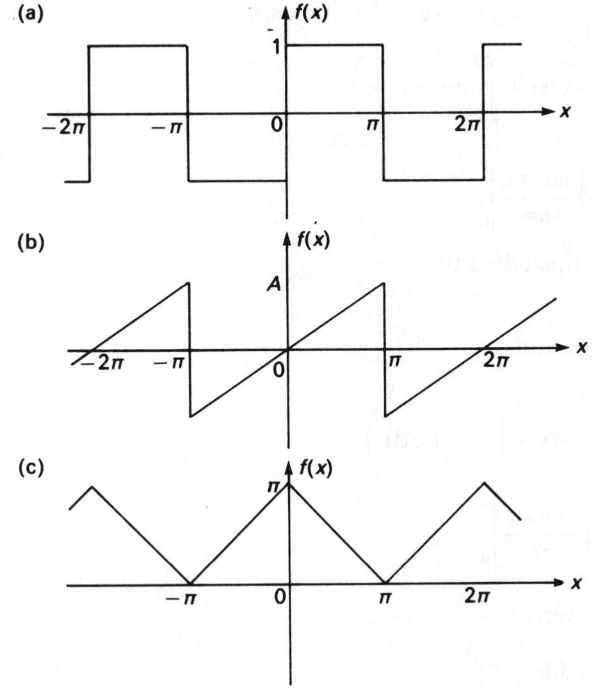

Figure 15.2

The restrictions on $f(x)$ are that, in the range $-\pi \leqslant x \leqslant \pi$, it must remain finite and single-valued (e.g. if it includes terms such as $\sqrt{x}$ then only the positive or only the negative root can be used), it must have a finite number of discontinuities, a finite number of maxima and minima, and some other features of a similar kind. Also, the periodicity must continue over the whole extent of the independent variable, from $-\infty$ to $+\infty$. The restrictions are very unlikely to cause you difficulty in the ordinary circumstances found with functions arising from real-world conditions, which don't contain infinities.

Example 15(a) Find the Fourier series expansion of the square-wave function shown in Fig. 15.2(a):

$$f(x)=\begin{cases}-1 & \text{for } -\pi \leqslant x \leqslant 0\\ +1 & \text{for } 0<x\leqslant \pi.\end{cases}$$

Answer: Use equations (15.5) and (15.6) to give

$$a_n=\frac{1}{\pi}\int_{-\pi}^{\pi}(\text{sign } x)\cos nx\,\mathrm{d}x \qquad \text{where (sign } x)=-1 \text{ if } -\pi \leqslant x \leqslant 0$$

$$\text{or } +1 \text{ if } 0<x\leqslant\pi$$

$$=\frac{1}{\pi}\left(-\int_{-\pi}^{0}\cos nx\,\mathrm{d}x+\int_{0}^{\pi}\cos nx\,\mathrm{d}x\right)$$

$$=\left[\frac{-\sin nx}{n\pi}\right]_{\pi}^{0}+\left[\frac{\sin nx}{n\pi}\right]_{0}^{\pi}$$

$$=0 \qquad \text{for all } n \text{ (including 0)}$$

$$b_n=\frac{1}{\pi}\int_{-\pi}^{\pi}(\text{sign } x)\sin nx\,\mathrm{d}x$$

$$=\frac{1}{\pi}\left(-\int_{-\pi}^{0}\sin nx\,\mathrm{d}x+\int_{0}^{\pi}\sin nx\,\mathrm{d}x\right)$$

$$=\left[\frac{\cos nx}{n\pi}\right]_{-\pi}^{0}-\left[\frac{\cos nx}{n\pi}\right]_{0}^{\pi}$$

$$=\begin{cases}0 & \text{for } n \text{ even}\\ \dfrac{4}{n\pi} & \text{for } n \text{ odd.}\end{cases}$$

$$\therefore \qquad f(x)=\frac{4}{\pi}\left(\frac{\sin x}{1}+\frac{\sin 3x}{3}+\frac{\sin 5x}{5}+\cdots\right).$$

Example 15(b) Find the Fourier series expansion of the sawtooth-wave function, shown in Fig. 15.2(b), with period 2π, given by $f(x)=Ax$ for $-\pi \leqslant x \leqslant \pi$.

Answer: For both a_n and b_n, integrate by parts.

$$\begin{aligned}
a_n &= \frac{1}{\pi}\int_{-\pi}^{\pi} Ax\cos nx\,\mathrm{d}x \\
&= \frac{A}{\pi}\left\{\left[x\frac{\sin nx}{n}\right]_{-\pi}^{\pi} - \int_{-\pi}^{\pi}\frac{\sin nx}{n}\,\mathrm{d}x\right\} \\
&= \frac{A}{\pi}\left\{0+\left[\frac{\cos nx}{n^2}\right]_{-\pi}^{\pi}\right\} \\
&= 0.
\end{aligned}$$

$$\begin{aligned}
b_n &= \frac{1}{\pi}\int_{-\pi}^{\pi} Ax\sin nx\,\mathrm{d}x \\
&= \frac{A}{\pi}\left\{\left[-x\frac{\cos nx}{n}\right]_{-\pi}^{\pi} + \int_{-\pi}^{\pi}\frac{\cos nx}{n}\,\mathrm{d}x\right\} \\
&= \frac{A}{\pi}\left\{\left[-x\frac{\cos nx}{n^2}\right]_{-\pi}^{\pi} + \left[\frac{\sin nx}{n^2}\right]_{-\pi}^{\pi}\right\},
\end{aligned}$$

but for $x=\pm\pi$,

$$\sin nx = 0 \text{ for all integers } n$$

$$\cos nx = \begin{cases} -1 & \text{for } n \text{ odd,} \\ +1 & \text{for } n \text{ even.} \end{cases}$$

$\therefore$ for n odd, $b_n = 2A/n$ and for n even, $b_n = -2A/n$. Finally,

$$a_0 = \frac{1}{2\pi}\int_{-\pi}^{\pi} Ax\,\mathrm{d}x = \frac{1}{4\pi}[Ax^2]_{-\pi}^{\pi} = 0.$$

Hence

$$f(x)=2A\left(\frac{\sin x}{1}-\frac{\sin 2x}{2}+\frac{\sin 3x}{3}-\frac{\sin 4x}{4}+\cdots\right).$$

Example 15(c) Find the Fourier series expansion for the triangular wave function, shown in Fig. 15.2(c), with period 2π, given by

$$f(x)=\begin{cases}\pi+x & \text{for } -\pi\leqslant x<0\\ \pi-x & \text{for } 0\leqslant x\leqslant\pi.\end{cases}$$

Answer: The integrals $\int_{-\pi}^{\pi} x \sin nx\, dx$ and $\int_{-\pi}^{\pi} x \cos nx\, dx$ were evaluated in example 15(b).

$$a_n=\frac{1}{\pi}\int_{-\pi}^{0}(\pi+x)\cos nx\,dx+\frac{1}{\pi}\int_{0}^{\pi}(\pi-x)\cos nx\,dx$$

$$=\frac{1}{\pi}\left[\frac{\pi}{n}\sin nx\right]_{-\pi}^{\pi}+\frac{1}{\pi}\left[x\frac{\sin nx}{n}\right]_{-\pi}^{0}+\frac{1}{\pi}\left[\frac{\cos nx}{n^2}\right]_{-\pi}^{0}$$

$$-\frac{1}{\pi}\left[x\frac{\sin nx}{n}\right]_{0}^{\pi}-\frac{1}{\pi}\left[\frac{\cos nx}{n^2}\right]_{0}^{\pi}.$$

For n odd,

$$a_n=0+0+\frac{2}{\pi n^2}+0-\left(-\frac{2}{\pi n^2}\right)$$

$$=\frac{4}{\pi n^2}.$$

For n even, $a_n=0$.

$$b_n=\frac{1}{\pi}\int_{-\pi}^{0}(\pi+x)\sin nx\,dx+\frac{1}{\pi}\int_{0}^{\pi}(\pi-x)\sin nx\,dx$$

$$=\frac{1}{\pi}\left[-\frac{\pi}{n}\cos nx\right]_{-\pi}^{\pi}-\frac{1}{\pi}\left[x\frac{\cos nx}{n}\right]_{-\pi}^{0}+\frac{1}{\pi}\left[\frac{\sin nx}{n^2}\right]_{-\pi}^{0}$$

$$+\frac{1}{\pi}\left[x\frac{\cos nx}{n}\right]_{0}^{\pi}-\frac{1}{\pi}\left[\frac{\sin nx}{n^2}\right]_{0}^{\pi}.$$

For n odd,

$$b_n=0+\frac{1}{n}+0-\frac{1}{n}-0=0.$$

For n even,

$$b_n=0-\frac{1}{n}+0+\frac{1}{n}+0=0.$$

$$a_0=\frac{1}{2\pi}\int_{-\pi}^{0}(\pi+x)\,dx+\frac{1}{2\pi}\int_{0}^{\pi}(\pi-x)\,dx$$

$$=\frac{1}{2\pi}\{[\pi x]_{-\pi}^{\pi}+[x^2/2]_{-\pi}^{0}-[x^2/2]_{0}^{\pi}\}$$

$$=0.$$

Hence

$$f(x)=\frac{4}{\pi}\left(\cos x+\frac{\cos 3x}{9}+\frac{\cos 5x}{25}+\cdots\right).$$

15.3 FOURIER SERIES FOR ODD AND EVEN FUNCTIONS

We have met before in section 11.5 the concept of odd and even functions; if $f(-x)=f(x)$ the function f is said to be even and if $f(-x)=-f(x)$ it is said to be odd. On this basis the sine function is odd and the cosine function is even. Some functions, of course, are neither odd nor even; for example, if $f(x)=3x^2-2x$, then $f(1)=1$ and $f(-1)=5$. However, the square wave of example 15(a) is odd and so is the sawtooth wave of example 15(b). This feature of the functions can be seen from graphs (a) and (b) of Fig. 15.2, as well as from the definitions of the functions. For the answers to both examples 15(a) and 15(b), $a_n=0$ for all n; i.e., each of these two odd functions has a Fourier series comprising only sine terms, which are themselves odd functions. Correspondingly, the function of example 15(c) is an even function, as seen from Fig. 15.2 (c) and from its definition; its Fourier series has $b_n=0$ for all n and comprises only cosine terms, which are themselves even functions. These results illustrate (though they don't prove) the following general rule.

> The Fourier series of an even function is a sum of cosines only ($b_n=0$ for all n) and the Fourier series of an odd function is a sum of sines only ($a_n=0$ for all n).

It is always worth checking whether a function to be expanded as a Fourier series is either odd or even; if it is, half of the integrals can be struck out, without making detailed evaluation of them.

EXERCISES 15B

1. Sketch the function $f(x)$ and obtain its Fourier series expansion where

$$f(x)=\begin{cases}1 & \text{for } -\pi\leqslant x\leqslant -1 \text{ and for } 1\leqslant x\leqslant \pi\\ x^2 & \text{for } -1\leqslant x\leqslant 1.\end{cases}$$

2. Find the Fourier series expansion for the function

$$f(x)=e^x \qquad \text{for } -\pi \leqslant x \leqslant \pi.$$

(Hint: note the integrals of example 10(1) and exercises 10E, number 5.)

3. A crude approximation to the left ventricular pressure during a heart beat is given by the periodic function $P(t)$ of period 2π, where

$$P(t)=\begin{cases} a+b\sin^2\left(\dfrac{3t}{2}\right) & \text{for } 0 \leqslant t \leqslant \dfrac{2\pi}{3} \\ a & \text{for } \dfrac{2\pi}{3} < t < 2\pi. \end{cases}$$

The interval 2π represents the R–R interval of the heart action, mitral value closure occurs at $t=0$, and a and b have the values 5 mm Hg and 120 mm Hg, respectively. Sketch this function and re-express it as a symmetrical function, $S(t)$, in the interval $(-\pi, \pi)$. Find the constant term and the first four harmonics of the Fourier series expansion of $S(t)$.

15.4 ALTERNATIVE WAYS OF WRITING FOURIER SERIES

There are several ways of writing Fourier series, and forms other than that of equation (15.4) are sometimes best suited to particular problems. One way of condensing the series is to use the auxiliary angle introduced in section 4.9 which allows $a\sin\theta+b\cos\theta$ to be written as $\sqrt{a^2+b^2}\sin(\theta+\phi)$ where $\tan\phi=b/a$. Another is to use the relation $e^{i\theta}=\cos\theta+i\sin\theta$. The equivalent ways of writing a Fourier series expansion of a function are then

$$f(x)=a_0+\sum_{n=1}^{\infty}(a_n\cos nx+b_n\sin nx) \tag{15.4}$$

$$=\sum_{n=0}^{\infty}(a_n\cos nx+b_n\sin nx) \tag{15.8}$$

$$=\sum_{n=0}^{\infty}c_n\cos(nx+\phi_n) \tag{15.9}$$

$$=\sum_{n=0}^{\infty}d_n\sin(nx+\theta_n) \tag{15.10}$$

$$=\sum_{n=-\infty}^{+\infty}g_n e^{inx} \tag{15.11}$$

Equation (15.11) follows from equation (15.4) because $\cos nx=\frac{1}{2}(e^{inx}+e^{-inx})$, etc. When the expansion is in the form of equation (15.11) it can easily be shown (provided i is treated just like any other number) that

$$g_n = \frac{1}{2\pi}\int_{-\pi}^{\pi} f(x)\,\mathrm{e}^{-inx}\,\mathrm{d}x. \tag{15.12}$$

In the above equations, a_n, b_n, c_n, d_n, and g_n are unknown amplitudes (of which g_n is usually complex, but the others are always real if $f(x)$ is real) and ϕ_n and θ_n are unknown phases. Since equations (15.4) and (15.8)–(15.11) are all equivalent, there are necessary relationships between the coefficients:

$$a_0 = g_0 = \text{mean (i.e. average) value of } f(x)$$

$$a_n = g_n + g_{-n} = c_n \cos\phi_n = d_n \sin\theta_n \qquad (n>1)$$

$$b_n = \mathrm{i}(g_n - g_{-n}) = -c_n \sin\phi_n = d_n \cos\theta_n \qquad (n>1)$$

$$g_n = \tfrac{1}{2}(a_n - \mathrm{i}b_n) \tag{15.13}$$

$$g_{-n} = \tfrac{1}{2}(a_n + \mathrm{i}b_n) \tag{15.14}$$

$$c_n^2 = d_n^2 = a_n^2 + b_n^2 = \pm 4g_n g_{-n} \qquad (n>1)$$

$$\tan\phi_n = -b_n/a_n \qquad (n>1)$$

$$\tan\theta_n = a_n/b_n \qquad (n>1).$$

If, as has been implicitly supposed, $f(x)$ is real, there are further simplifications:

$$g_n = g_{-n}^* \qquad \text{(i.e. } g_n \text{ and } g_{-n} \text{ are complex conjugates)}$$

$$a_n = \mathscr{R}(2g_n)$$

$$b_n = -\mathscr{I}(2g_n)$$

where $\mathscr{R}$ and $\mathscr{I}$ stand respectively for 'real part of' and 'imaginary part of' the number indicated, as in section 5.3.

If $f(x)$ is an even function of x it has been noted that $b_n=0$ for all n, but also in this case g_n is real and $g_n = g_{-n}$, $\phi_n=0$, $\theta_n=\pi/2$, $a_0=g_0$ and $a_n = c_n = d_n = 2g_n$.

Correspondingly, if $f(x)$ is odd it has been noted that $a_n=0$ for all n, but also in this case g_n is pure imaginary and equal to $-g_{-n}$, $\phi_n=\pi/2$, $\theta_n=0$ and $b_n=c_n=d_n=2\mathrm{i}g_n$.

Example 15(d) Find the Fourier expansion of the triangular wave of example 15(c) in the complex exponential form, i.e. find

$$f(\omega t) = \sum_{-\infty}^{\infty} g_k\,\mathrm{e}^{ik\omega t}$$

where

$$f(\omega t) = \begin{cases} \pi + \omega t & \text{for } -\pi \leqslant \omega t \leqslant 0 \\ \pi - \omega t & \text{for } 0 < \omega t \leqslant \pi. \end{cases}$$

Answer: To find the expansion in this case, it is simplest to treat ωt as the variable, rather than t itself. For convenience, look first at the integrals which will appear in the working out of the expansion. Note that i is treated as if it were an ordinary constant number.

1. $$\int_a^b e^{-ik\omega t}\, d(\omega t) = -\frac{1}{ik}\left[e^{-ik\omega t}\right]_a^b$$
$$= -\frac{1}{ik}\left(e^{-ikb} - e^{-ika}\right).$$

Now, $e^{-\theta} = \cos\theta - i\sin\theta$. Hence for $\theta = \pm\pi$,

$$e^{-ik\theta} = \begin{cases} 1 & \text{for } k \text{ even} \\ -1 & \text{for } k \text{ odd.} \end{cases}$$

Thus

$$\int_{-\pi}^{\pi} e^{-ik\omega t}\, d(\omega t) = 0$$

for all integer values of k.

2. $$\int_a^b \omega t\, e^{-ik\omega t}\, d(\omega t) = \left[\frac{-\omega t\, e^{-ik\omega t}}{ik}\right]_a^b + \frac{1}{ik}\int_a^b e^{-ik\omega t}\, d(\omega t)$$
$$= \left[\frac{-\omega t\, e^{-ik\omega t}}{ik}\right]_a^b + \frac{1}{k^2}\left[e^{-ik\omega t}\right]_a^b$$
$$= \left(\frac{e^{-ikb}}{k^2} - \frac{be^{-ikb}}{ik}\right) - \left(\frac{e^{-ika}}{k^2} - \frac{ae^{-ika}}{ik}\right).$$

These integrals can now be applied to the given problem.

$$2\pi g_n = \int_{-\pi}^{0} (\pi + \omega t)\, e^{-in\omega t}\, d(\omega t) + \int_0^{\pi} (\pi - \omega t)\, e^{-in\omega t}\, d(\omega t)$$

But

$$\int_{-\pi}^{0} \pi\, e^{-in\omega t}\, d(\omega t) + \int_0^{\pi} \pi\, e^{-in\omega t}\, d(\omega t) = \pi\int_{-\pi}^{\pi} e^{-in\omega t}\, d(\omega t)$$
$$= 0 \text{ from 1. for all } n.$$

$$\therefore \quad 2\pi g_n = \int_{-\pi}^{0} \omega t\, e^{-in\omega t}\, d(\omega t) - \int_{\theta}^{\pi} \omega t\, e^{-in\omega t}\, d(\omega t)$$
$$= \left[-\frac{\omega t}{in} e^{-in\omega t}\right]_{-\pi}^{0} + \left[\frac{1}{n^2} e^{-in\omega t}\right]_{-\pi}^{0}$$
$$- \left[-\frac{\omega t}{in} e^{-in\omega t}\right]_{0}^{\pi} - \left[\frac{1}{n^2} e^{-in\omega t}\right]_{0}^{\pi}.$$

Then, for n odd,

$$2\pi g_n=\left(0+\frac{\pi}{in}\right)+\left(\frac{1}{n^2}+\frac{1}{n^2}\right)-\left(\frac{\pi}{in}-0\right)-\left(-\frac{1}{n^2}-\frac{1}{n^2}\right),$$

$$=\frac{4}{n^2}$$

and for n even

$$2\pi g_n=\left(0-\frac{\pi}{in}\right)+\left(\frac{1}{n^2}-\frac{1}{n^2}\right)-\left(\frac{\pi}{in}-0\right)-\left(\frac{1}{n^2}-\frac{1}{n^2}\right)$$

$$=0.$$

Hence

$$g_n=\frac{2}{\pi n^2} \qquad \text{(for } n \text{ odd)}$$

$$=g_{-n}, \qquad \text{since } n \text{ appears squared.}$$

Thus finally

$$f(\omega t)=\sum_{-\infty}^{\infty}\frac{2}{\pi}\frac{e^{-in\omega t}}{n^2} \qquad \text{(for } n \text{ odd)} \qquad (15.15)$$

Equation (15.15) could also be written as

$$f(\omega t)=\sum_{1}^{\infty}\frac{2}{\pi n^2}(e^{in\omega t}+e^{-in\omega t}) \qquad \text{since } g_n=g_{-n}$$

$$=\frac{4}{\pi}\sum_{1}^{\infty}\frac{\cos n\omega t}{n^2} \qquad \text{(for } n \text{ odd).} \qquad (15.16)$$

The last version, equation (15.16), is in the same form as the answer given for example 15(c) as is to be expected. Conversion between the various forms follows from equations (15.13) and (15.14) when $b_n=0$.

EXERCISE 15C

Either directly, by use of equations (15.11) and (15.12), or indirectly, by use of equations (15.13) and (15.14) and related equations, express as imaginary exponential Fourier series the functions of exercises 15B, viz.

(a) $f(x)=\begin{cases}1 & \text{for } -\pi\leqslant x<-1 \text{ and } 1<x\leqslant\pi \\ x^2 & \text{for } -1\leqslant x\leqslant 1;\end{cases}$

(b) $f(x)=e^x \quad \text{for } -\pi\leqslant x\leqslant\pi.$

15.5 CHANGE OF SCALE

The quantity 2π is not the only possible interval of periodicity. Suppose that $f(x)$ is periodic with a period $2L$ rather than 2π. Then we can write

$$f(x)=a_0+\sum_{m=1}^{\infty}\left(a_m\cos\frac{m\pi x}{L}+b_m\sin\frac{m\pi x}{L}\right). \tag{15.17}$$

The sets of orthogonal functions are now $\cos(m\pi x/L)$ and $\sin(m\pi x/L)$ in the range $-L$ to L for $m=1, 2, 3,\ldots$ It is then easily shown, by arguments exactly corresponding to those used before, that the coefficients in equation (15.17) are given by

$$a_m=\frac{1}{L}\int_{-L}^{L} f(x)\cos\frac{m\pi x}{L}\,dx$$

$$b_m=\frac{1}{L}\int_{-L}^{L} f(x)\sin\frac{m\pi x}{L}\,dx$$

$$a_0=\frac{1}{2L}\int_{-L}^{L} f(x)\,dx.$$

Alternatively, using the exponential function, we would have

$$f(x)=\sum_{-\infty}^{\infty} g_m\,e^{-im\pi x/L}$$

where

$$g_m=\frac{1}{2L}\int_{-L}^{L} f(x)\,e^{-im\pi x/L}\,dx.$$

Example 15(e) Find the Fourier series expansion of the function $f(t)$ which is periodic in the range $-L$ to L and is defined by

$$f(t)=\begin{cases}1 & \text{for } -p\leqslant t\leqslant p\\ 0 & \text{for } -L\leqslant t<-p \text{ and } p<t\leqslant L.\end{cases}$$

Answer: The function is a train of pulses of length $2p$ with centres separated by $2L$, as shown in Fig. 15.3. Furthermore, the function is even, so $b_n=0$ for all n. The integral for a_n runs only from $-p$ to p since the function has zero value in the remainder of the period. Hence

$$a_n=\frac{1}{L}\int_{-p}^{p}\cos\frac{n\pi t}{L}\,dt$$

$$=\frac{1}{L}\left[\frac{L}{n\pi}\sin\frac{n\pi t}{L}\right]_{-p}^{p}$$

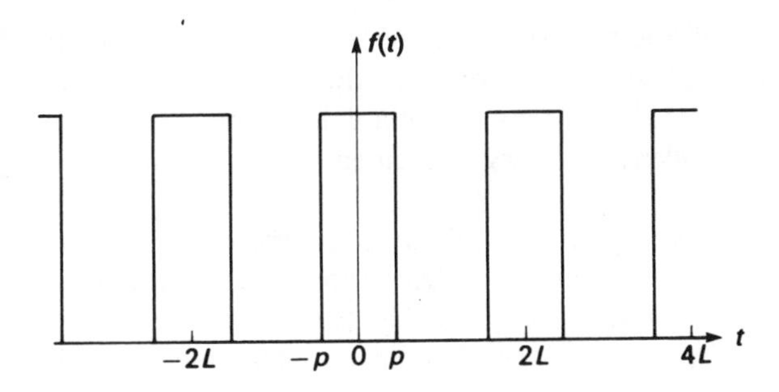

Figure 15.3

$$=\frac{1}{L}\left(\frac{2L}{n\pi}\sin\frac{n\pi p}{L}\right)$$

$$=\frac{2}{n\pi}\sin\frac{n\pi p}{L}.$$

$$a_0=\frac{1}{2L}\int_{-p}^{p}1\cdot dx=\frac{2p}{2L}=\frac{p}{L}.$$

Thus

$$f(t)=\frac{2}{\pi}\left(\frac{p\pi}{2L}+\frac{1}{1}\sin\frac{\pi p}{L}\cos\frac{\pi t}{L}+\frac{1}{2}\sin\frac{2\pi p}{L}\cos\frac{2\pi t}{L}+\cdots\right) \tag{15.18}$$

EXERCISE 15D

Find the Fourier series expansion for the function $f(t)$ with period $2Q$ where

$$f(t)=\begin{cases} p+t & \text{for } -p<t<0 \\ p-t & \text{for } 0<t<p \quad \text{(where } p<Q) \\ 0 & \text{otherwise, in the period } -Q \text{ to } Q. \end{cases}$$

15.6 FROM FOURIER SERIES TO FOURIER TRANSFORM

In practice there are two severe difficulties in relying on Fourier series for the description of oscillating or periodic phenomena.

The first is the requirement of the theory that the oscillations have been going on since time and space began and will continue for ever. Obviously, the oscillations we actually meet don't do that. The heart which produced the ECG signal shown in Fig. 15.1 may possibly, if its owner doesn't smoke or drink too much, last out for between two and three thousand million beats; but not for ever. Even the periodicity of the Moon had a beginning and will come to an end; and in comparison with these, most oscillations we meet are pathetically short-lived: tens of second for a plucked guitar string, only one or two wavelengths of leg movement along a centipede's body.

But the second difficulty is worse: most of the phenomena we want to study aren't periodic at all. There is no periodicity about a heart attack, or the transformation of a tadpole into a frog, or the structure of the protein molecule lysozyme. And even many of the phenomena which may initially seem to be periodic are only quasi-periodic when examined in detail. Heartbeats, like those leading to Fig. 15.1, are found to have a variable interbeat interval, depending on breathing, emotional response, digestion, etc. And the hen which lays an egg a day doesn't do so to the minute.

For dealing with these limitations to the applications of Fourier series, the theory is extended into what is called the Fourier transform. The Fourier series provides a means of studying the frequencies in a repetitive event; the Fourier transform in effect provides a means of studying the frequencies inherent in a single isolated event.

We are concerned here to make a beginning of a study of Fourier transforms. What we shall do is to stretch out the interpulse period of a Fourier series like that developed for the function graphed in Fig. 15.3. This stretching is done leaving just one oscillation centre stage and allowing all the others to exit stage left or stage right in such a way that the Fourier series formula in the complex exponential form, equation (15.11), is changed into the Fourier transform formula. After this has been done we shall be able to look back and see what it means.

15.7 THE FOURIER TRANSFORM

Take, as the example of the Fourier series to be sex-changed, the series for the train of rectangular pulses in the function examined in example 15(e). Start by writing ω_0 for π/L. Then ω_0 is the fundamental frequency of the Fourier series expansion, because substituting it into equation (15.18) gives

$$f(t)=\frac{2L\omega_0}{\pi^2}\left\{\frac{p\omega_0}{2}+\frac{\sin\omega_0 p}{1}\cos\omega_0 t+\frac{\sin 2\omega_0 p}{2}\cos 2\omega_0 t\right.$$
$$\left.+\frac{\sin 3\omega_0 p}{3}\cos 3\omega_0 t+\cdots\right\}$$

Thus ω_0 is the (circular) frequency of the first oscillatory term. The other terms have frequencies which are harmonics (or multiples) of this frequency. Now, $2L$ is the interval between one pulse and the next. As L becomes greater and greater, we retain the central pulse which lasts from $-p$ to p, but the gaps between it and the one before and the one after become indefinitely large. Also, as L becomes very large, $\omega_0 = \pi/L$ becomes very small. Indeed, as $L \to \infty$ so $\omega_0 \to 0$, although $\omega_0 L$ remains constant and equal to π. Furthermore, the harmonics of the Fourier series expansion crowd up on each other: the separation of the $(n+1)$th harmonic from the nth is ω_0 and as ω_0 shrinks right down we replace it by $\delta\omega$ to stress that there is an infinitesimally small frequency separation between the harmonics.

With L tending towards infinity, we now write $n\omega_0 = \omega$. As ω_0 tends to zero, we must go to very high harmonics – very large values of n – to have a product of appreciable value, and as there will then be only the infinitesimally small change of ω between $n\omega_0$ and $(n+1)\omega_0$, we can consider $\omega = n\omega_0$ as a continuous variable.

The original Fourier series could have been written (cf. equation (15.11) as

$$f(t) = \sum_{-\infty}^{\infty} g_n \, e^{in\omega_0 t} \tag{15.19}$$

where

$$g_n = \frac{1}{2L} \int_{-L}^{L} f(t) \, e^{-in\omega_0 t} \, dt. \tag{15.20}$$

But now, as $L \to \infty$, if we write $2Lg_n = F$, equation (15.20) can be written as

$$F = \int_{-L}^{L} f(t) \, e^{-in\omega_0 t} \, dt.$$

Since F depends on $n\omega_0$ which is to be replaced by the continuous variable ω, F can be considered as a function of ω defined by

$$F(\omega) = \lim_{L \to \infty} \int_{-L}^{L} f(t) \, e^{-in\omega_0 t} \, dt,$$

i.e.

$$F(\omega) = \int_{-\infty}^{\infty} f(t) \, e^{-i\omega t} \, dt \tag{15.21}$$

At the same time we have, by considering what happens to equation (15.19),

$$f(t) = \lim_{L \to \infty} \sum g_n \, e^{in\omega_0 t}$$

$$= \lim_{L \to \infty} \frac{1}{2\pi} \sum (2Lg_n) \, e^{in\omega_0 t} \frac{\pi}{L}.$$

Since $\pi/L = \omega_0$ which becomes $\delta\omega$ as L becomes very large, if we go to the limit of $L\to\infty$ the summation becomes an integral and

$$f(t)=\frac{1}{2\pi}\int_{-\infty}^{\infty} F(\omega)\,e^{i\omega t}\,d\omega. \tag{15.22}$$

The two equations, equations (15.21) and (15.22) are very similar; the two functions $f(t)$ and $F(\omega)$ are said to form a Fourier pair, sometimes written as $f(t)\Leftrightarrow F(\omega)$. The function $F(\omega)$ is the Fourier transform of $f(t)$, given by the so-called forward transformation of equation (15.21) and the function $f(t)$ is the Fourier transform of $F(\omega)$, given by the reverse transformation of equation (15.22).

15.8 EXAMPLE OF A FOURIER TRANSFORM

As an example of the calculation of a Fourier transform, we can look at the problem presented by one pulse as illustrated in Fig. 15.4 – not a drum roll as in Fig. 15.3 which could be expressed as a Fourier series, but just one beat of the drum or a snap of the fingers. The function describing it is $f(t)$ where

$$f(t)=\begin{cases}0 & \text{for } t<-a \text{ and for } t>a\\ A & \text{for } -a\leqslant t\leqslant a.\end{cases}$$

Then

$$\begin{aligned}F(\omega)&=\int_{-\infty}^{\infty} f(t)\,e^{-i\omega t}\,dt\\ &=A\int_{-a}^{a} e^{-i\omega t}\,dt\\ &=\frac{A}{i\omega}(e^{-i\omega a}-e^{-i\omega a})\\ &=2A\frac{\sin\omega a}{\omega}.\end{aligned} \tag{15.23}$$

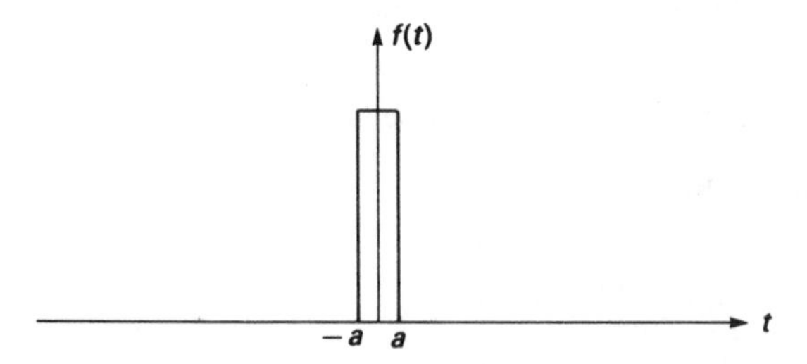

Figure 15.4

The reverse transform, given by equation (15.22) is

$$f(t)=\frac{2A}{2\pi}\int_{-\infty}^{\infty}\frac{\sin a\omega}{\omega}\,e^{i\omega t}\,d\omega.$$

This is simple to write down, but complicated to evaluate, so I shall omit its evaluation here. It does, however, lead back to $f(t)$ as originally defined.

In the case of a signal, like that of the ECG record in Fig. 15.1(b), which cannot be given a mathematical description, it is possible to construct the Fourier transform out of a series of samples of the signal. Thus, measurement of the ECG signal at millisecond intervals might be used to make a Fourier analysis of heart action. However, the deployment of these techniques depends on computer systems and signal processing methods, as well as on the theory of Fourier transforms, and goes beyond the scope of this text.

EXERCISE 15E

Find the Fourier transform of a single triangular pulse like one of those in exercises 15D, number 1, i.e.

$$f(t)=\begin{cases} p+t & \text{for } -p\leqslant t\leqslant 0 \\ p-t & \text{for } 0<t\leqslant p \\ 0 & \text{for } t<-p \text{ and for } t>p. \end{cases}$$

15.9 WHAT IS A FOURIER TRANSFORM?

Here is one way (the fairly conventional way) of looking at the matter. In the case of the Fourier series for a periodic pulse of width $2p$ and period $2L$, given by equation (15.18), the successive coefficients initially reduce in size. Consider, for example, the duration and spacing of the pulses when $p=1$ and $L=16$. The successive coefficients are

$$\frac{\sin(\pi/16)}{1},\ \frac{\sin(2\pi/16)}{2},\ \frac{\sin(3\pi/16)}{3},\ldots$$

Evidently the 16th coefficient will be zero; after that there are 15 negative coefficients, a zero coefficient, then another 15 positive coefficients, and so on. The size of the coefficients, plotted against np/L, gives the diagram of Fig. 15.5. The separation between the lines of Fig. 15.5 is p/L and the first zero occurs when $np/L=1$, i.e. when $n=L/p$. If now L is increased, as a move towards turning the Fourier series into a Fourier transform, but p is unchanged, the positions of the zeros in Fig. 15.5 remain at the same place on the abscissa (namely, at $np/L=1, 2, 3, \ldots$) but there are more lines to be

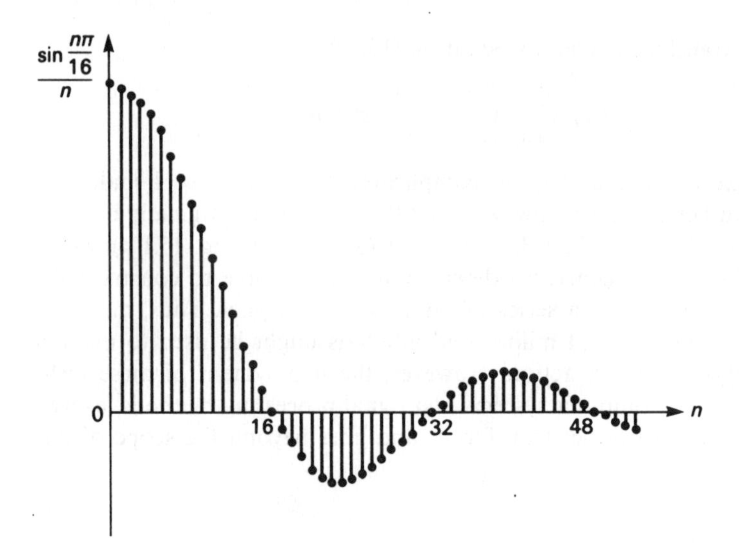

Figure 15.5

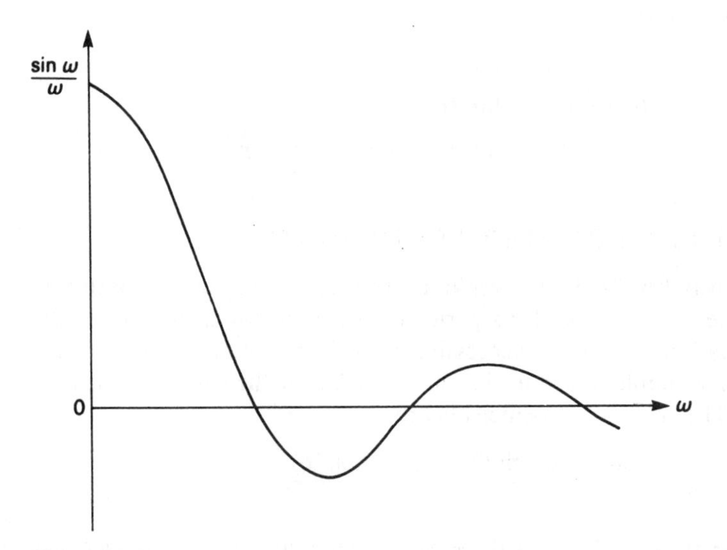

Figure 15.6

fitted in, so they become more densely spaced. The Fourier transform is this process pushed to the limit. Fig. 15.6 is a graph of $F(\omega)$ from equation (15.23). It can be considered as the envelope of the 'stick graph' of Fig. 15.5, the graph obtained by joining the tops of the lines representing the coefficients in the Fourier series for well-separated pulses.

Thus the Fourier transform is to a Fourier series something like a continuous statistical distribution to a discrete one. A repetitious signal can be represented by a set of separated oscillations, the fundamental frequency and its harmonics. But the fewer the repetitions the more frequencies must be brought in, and a single isolated event is as if created by the combined activity of oscillators of every possible frequency from zero to infinity.

But why an infinite spectrum of frequencies? The reason that the very highest frequencies are required is connected with clarity of detail. The ceasing of a flute is a sharper event than the ceasing of the bass notes of an organ because the flute voice registers at a so much higher frequency. If a painter wishes to include the finest detail, the minutest brush strokes are necessary because these involve the highest (spatial) frequencies. The pulse of Fig. 15.4 begins and ends with extreme suddenness, as also do the pulses shown in Fig. 15.3, and to catch the sharpness of the corners of the pulse requires very high frequencies – in fact, infinitely high, though they are of low amplitude. The continuity of the spectrum of frequencies is required in the Fourier transform in order to suppress echoes – occasional juxtapositions of wave peaks which would provide other pulses – the other pulses of the repetitious signal of Fig. 15.3.

As another way of looking at the matter, suppose that the pulse of Fig. 15.4 represents a sound in time. (It needn't be: it could represent a brick lying on a flat surface, and so be a mass in space; or it could be the amount of vibration in a car as a function of the speed of the vehicle.) The requirement is to describe the sound as it changes in comparison with time as the clock progresses, and we are so used to devices which describe time as changing very steadily that we can hardly think of it in any other way. But there are, nevertheless, other ways and other measures. Personal biological time goes forward in fits and starts; sometimes it slows, sometimes it races. Because it is so individual, we can't properly share it with others and therefore use clocks as a common reference and measure. However, there are measures of time which differ from the records of clocks, and frequency is one such. Frequency is measured in reciprocal time so frequency could be used to mark the passage of time. As a fanciful example, suppose that along a pair of railway tracks, for some part of the day, trains went in one direction at a regular rate (frequency) of three an hour and in the other direction at a regular rate of four an hour. Then there would be a number of places at which trains passed each other, going in opposite directions. If you stood at such a place you could mark the passage of hours by those moments when the trains passed – those moments, occurring precisely once an hour, when the two frequencies were precisely in phase. The more frequencies you encompassed in your system, the finer you could subdivide the passage of time.

According to this last viewpoint, a Fourier transform is merely a method of measuring an event (occurring in time or space or temperature or whatever) by a different measuring system. The users of Fourier transforms (abbreviated as FTs) speak of operating in either the time (or space, etc.) domain or the frequency domain. There is no Berlin wall between these two domains; $f(t)$ and $F(\omega)$ are totally interchangeable by means of the operations of equations (15.21) and (15.22). A function and its FT contain identical information, in the same way that a single quantity is expressed either by an ordinary number or by its logarithm to a stated base. By the use of FTs, a problem may be looked at from two points of view and the simpler way of solving it can be found in the more appropriate domain.

15.10 THE PURPOSE OF TRANSFORM MATHEMATICS

It may be that the idea of changing the vantage point to look at a mathematical problem seems to you an improbable way of getting a better view of it. If so, consider the following example.

Example 15(f) Multiply XXIV by CLIII.

Answer: First transform the roman numerals into arabic ones with decimal notation. This presents the problem as the calculation of 24×153. Within your electronic calculator, transform the arabic numerals into binary ones, multiply, and transform back into arabic numerals to get 3672, and finally transform into roman numerals to get MMMDCLXXII.

Not a fair example, you say. Well, yes and no. The Romans got on well enough at addition, but weren't very good at multiplication. Going from roman to arabic numerals doesn't change the way of counting, but it does switch the mode of expressing numbers into a format which allows complicated multiplication to be efficiently accomplished by use of a set of simple rules.

In a comparable way, switching the description of a process from one in the time or space domain to one in the frequency domain allows more efficient computation of certain problems. For example, the action of capacitances and inductances in electronics depends directly on frequency (and so only inversely on time). Hence a frequency-domain description of certain properties of electronic circuits allows the effects of these components to be easily calculated. The same is true of viscous and elastic elements in models which are formulated of muscle function. Of course, some aspects of muscle activity (and also some aspects of the activity of electronic circuits) take place directly in the time domain and can best be studied without recourse to Fourier transforms. Each problem, and sometimes the separate fractions of a problem, must be assessed for the route to the easiest method of solution and can then be picked off by applying that method. Even if the computation of Fourier transforms may seem complicated, the calculation of the solution to certain problems, while avoiding the employment of Fourier transforms, would be much more complicated still.

> *... we should, with joy, revel,*
> *pleasance, and applause, transform*
> *ourselves ...*
>
> William Shakespeare, *Othello*

Answers to exercises

In most cases, the problems set in the exercises are solved by the same method as is illustrated in the examples immediately preceding the exercise.

EXERCISES 2A

1. (a) 85; (b) 22737; (c) 703710.
2. 1100001 (base 2).
3. 451 (base 8).
4. 11c4f (base 16).

EXERCISES 2B

1. $(a, b)\cdot(1, 1)=(a\cdot 1, b\cdot 1)=(a, b)$;
   $(a, b)/(1, 1)=(a/1, b/1)=(a, b)$;
   $(a, b)+(0, k)=(ak+0, bk)=(a, b)$;
   $(a, b)\cdot(0, k)=(0, bk)=(0, k)$;
   hence $(1, 1)$ and $(0, k)$ are the unity and zero terms of the system.
2. (a) $(12, 7)$; (b) $(-9, 0)$; (c) $(a, b)\cdot(1, 0)=(a-0, 0+\mathrm{b})=(a, b)$ so that $(1, 0)$ is the unity term of the system.

EXERCISES 3A

1. (a) 27/64; (b) $\sqrt[4]{27}/8$; (c) 8.
2. The expression is
$$\begin{aligned}(a+2b)(3(a+2b)^2-2(a+2b)-1 &= (a+2b)(3(a+2\mathrm{b})+1)((a+2b)-1)\\ &= (a+2b)(3a+6b+1)(a+2b-1).\end{aligned}$$

EXERCISES 3B

1. Seat the man leaving 2 possible places for the woman. There are then 3! ways of seating the other men and 3! ways of seating the other women. All other positions for the man will lead to the same arrangement round the circular table, so the total number of distinct possibilities is $2(3!)^2=72$.

2. Starting at A and finishing at F there are 4 towns in between, so 4! possible routes. Similarly starting at F and finishing at A gives 4! routes. Hence the total number of routes is $2 \cdot 4! = 48$.

3. Choose 5 from 24, i.e., $^{24}C_5 = \dfrac{24 \cdot 23 \cdot 22 \cdot 21 \cdot 20}{1 \cdot 2 \cdot 3 \cdot 4 \cdot 5} = 42504$ ways.

   (a) Choose n men from 9 and $5-n$ women from 15 when $n=1, 2, 3, 4$, or 5. Hence total number of ways is N where

$$N = {}^9C_1^{15}C_4 + {}^9C_2{}^{15}C_3 + {}^9C_3{}^{15}C_2 + {}^9C_4{}^{15}C_1 + {}^9C_5{}^{15}C_0$$
$$= 39501.$$

   (b) As in (a) but omitting the case of $n=5$, so $N' = 39375$.
   (c) As in (a) but with $n=0$, 1, or 2; so

$$N'' = {}^9C_0^{15}C_5 + {}^9C_1^{15}C_4 + {}^9C_2^{15}C_3 = 31668.$$

EXERCISES 3C

1. Suppose the formula is correct for n so

$$\sum_{i=1}^{n} i^2 = \frac{n(n+1)(2n+1)}{6}.$$

   In that case

$$\sum_{i=1}^{n+1} i^2 = \frac{n(n+1)(2n+1)}{6} + (n+1)^2$$
$$= \frac{(n+1)(n+2)(2n+3)}{6}.$$

   So if the formula is correct for n it is also correct for $n+1$. But it is correct for $n=1$ (and for $n=2$); hence it must be correct for all n.

2. Suppose the statement is true so that $n^3 - n = 3k$, where k is an integer. Then

$$(n+1)^3 - (n+1) = n^3 + 3n^2 + 3n + 1 - n - 1$$
$$= (n^3 - n) + 3(n^2 + n)$$
$$= 3k + 3(n^2 + n).$$

   Hence, if the statement is correct for n it is also correct for $n+1$. But it is correct for $n=1$ (and also for $n=2$); so it must be correct for all values of n.

EXERCISES 3D

1. $(p+q)^{16}=P(1+s)^{16}$ where $P=p^{16}=0.6^{16}$ and $s=q/p=2/3$. In $(1+s)^{16}$, let u_r be the rth term, etc. where $u_r={}^nC_{r-1}s^{r-1}$. Hence

$$u_{r+1}=\frac{n-(r-1)}{r}su_r$$

and $u_{r+1}>u_r$ if

$$\frac{16-r+1}{r}\cdot\frac{2}{3}\cdot u_r>u_r,$$

i.e. if $(34-2r)>3r$, i.e. if $5r<34$, i.e. if $r<6\frac{4}{5}$. But r must be an integer, so r must have the value 6 and the maximum term is $u_7={}^{16}C_6(2/3)^6\approx 703$. ($u_6\approx 575$; $u_8\approx 669$.) In $(p+q)^{16}$, with $p=0.6$, $q=0.4$, the largest term is $P\cdot u_7={}^{16}C_6p^{10}q^6=0.198$. In a test with 60% chance of success (p) and 40% of failure (q) the most probable outcome of 16 repeats is ten successes and six failures and the chance of this is 19.8%.

2. (a) $1+\frac{x}{2}-\frac{x^2}{8}+\frac{x^3}{16}-\frac{5x^4}{128}+\cdots$

(b) $1-6y^2+27y^4-108y^6+405y^8-\cdots$

(c) $1+z+z^2+z^3+z^4+\cdots$

3. $$\sqrt{\frac{1-x}{1+x}}=(1-x)^{\frac{1}{2}}(1+x)^{-\frac{1}{2}}=\left(1-\frac{x}{2}-\frac{x^2}{8}-\frac{x^3}{16}-\frac{5x^4}{128}\cdots\right)$$

$$\times\left(1-\frac{x}{2}+\frac{3x^2}{8}-\frac{5x^3}{16}+\frac{35x^4}{128}\cdots\right)$$

$$=1-x+\frac{x^2}{2}-\frac{x^3}{2}+\frac{3x^4}{8}+\cdots$$

If $x=0.15$, $\sqrt{(1-x)/(1+x)}=0.8597$; the first four terms give 0.8596.

EXERCISES 4A

1. (a) $\tan\theta+\cot\theta=\frac{\sin\theta}{\cos\theta}+\frac{\cos\theta}{\sin\theta}=\frac{\sin^2\theta+\cos^2\theta}{\sin\theta\cos\theta}=\frac{1}{\sin\theta\cos\theta}=\sec\theta\operatorname{cosec}\theta.$

(b) $\frac{\sin\theta}{1+\cos\theta}=\frac{\sin\theta(1-\cos\theta)}{1-\cos^2\theta}=\frac{1-\cos\theta}{\sin\theta}=\operatorname{cosec}\theta-\cot\theta.$

2. $$\frac{\sin^4\theta+\cos^4\theta}{\sin\theta\cos\theta}=\frac{(\sin^2\theta+\cos^2\theta)^2-2\sin^2\theta\cos^2\theta}{\sin\theta\cos\theta}=\frac{1-2\sin^2\theta\cos^2\theta}{\sin\theta\cos\theta}$$
$$=\sec\theta\,\text{cosec}\,\theta-2\sin\theta\cos\theta.$$

EXERCISES 4B

1. $$\cos 3X=\cos(X+2X)=\cos X\cos 2X-\sin X\sin 2X$$
$$=\cos X(2\cos^2 X-1)-\sin X(2\sin X\cos X)$$
$$=2\cos^3 X-\cos X-2\cos X(1-\cos^2 X)=4\cos^3 X-3\cos X.$$
If $3X=\pi/2$, $X=\pi/6$, $\cos 3X=0$, $\cos X=\sqrt{3}/2$.

2. $$\frac{1}{\cot\theta-\tan\theta}-\frac{1}{\cot\theta+\tan\theta}=\frac{2\tan\theta}{\cot^2\theta-\tan^2\theta}=\frac{2\sin^3\theta\cos\theta}{\cos^4\theta-\sin^4\theta}$$
$$=\frac{2\sin^3\theta\cos\theta}{(\cos^2\theta+\sin^2\theta)(\cos^2\theta-\sin^2\theta)}$$
$$=\frac{\sin 2\theta\sin^2\theta}{\cos 2\theta}=\frac{\tan 2\theta}{\text{cosec}^2\theta}.$$

EXERCISES 4C

1. $$\frac{\sin 6X+\sin 2X}{\cos 6X+\cos 2X}=\frac{2\sin 4X\cos 2X}{2\cos 4X\cos 2X}=\tan 4X.$$
2. (a) $\cos\pi/6=\sqrt{3}/2$. But $\cos\pi/6=2\cos^2\pi/12-1$. $\therefore \cos\pi/12=0.966$;
   (b) $\cos\pi/6=1-2\sin^2\pi/12$. $\therefore$ $\sin\pi/12=0.259$;
   (c) $\tan\pi/12=(\sin\pi/12)/(\cos\pi/12)=0.268$.

EXERCISES 4D

1. (a) $13\sin(A+C)$, i.e. $B=13$, $C=\tan^{-1}5/12=0.395$ radians;
   (b) $\sqrt{13}\sin(A+C)$, i.e. $B=\sqrt{13}$, $C=\tan^{-1}2/3=0.588$ radians.
2. $10\sin 2\pi ft+B\cos 2\pi ft=A\sin(2\pi ft+\phi)$, where $A=\sqrt{10^2+B^2}$ and $\tan\phi=B/10$. But if $\phi=\pi/3$, then $\tan\phi=\sqrt{3}$, so that $B=10\sqrt{3}$.

EXERCISES 5A

1. With $a=2$, $\dfrac{(1,0)}{(1,1)}=\dfrac{(1,0)}{(1,1)}\dfrac{(1,-1)}{(1,-1)}=\dfrac{(1,-1)}{(-1,0)}=(-1,1)$.

2. $(m+n\sqrt{a})^r = m^r + {}^rC_1 m^{r-1} n\sqrt{a} + {}^rC_2 m^{r-2} n^2 a + {}^rC_3 m^{r-3} n^3 a\sqrt{a} + \cdots$
$$= \{m^r + {}^rC_2 m^{r-2} n^2 a + {}^rC_4 m^{r-4} n^4 a^2 + {}^rC_6 m^{r-6} n^6 a^3 + \cdots\}$$
$$+ \{{}^rC_1 m^{r-1} n + {}^rC_3 m^{r-3} n^3 a + {}^rC_5 m^{r-5} n^5 a^2 + \cdots\}\sqrt{a}.$$
$(m+n\sqrt{a})^r = (A, B)$ in number pair notation, where A, B are the above brackets.

EXERCISES 5B

1. $z_1 + z_2 = 3p + iq$; $z_1 z_2 = 2(p^2+q^2) + i3pq$.
2. $(\cos A + i\sin A)^3 = \cos^3 A - 3\cos A \sin^2 A + i(3\cos^2 A \sin A - \sin^3 A)$. $\cos \pi/3 = 1/2$, $\sin \pi/3 = \sqrt{3}/2$. Substitute these values and the result follows.

EXERCISES 5C

1. $z^3 = \cos 3\theta + i\sin 3\theta$, $z^{-1} = \cos\theta - i\sin\theta$, $z^{-3} = \cos 3\theta - i\sin 3\theta$. Hence $z - z^{-1} = i2\sin\theta$ so $(z-z^{-1})^3 = (i2\sin\theta)^3 = -i8\sin^3\theta$. But also $(z-z^{-1})^3 = z^3 - 3z^2z^{-1} + 3zz^{-2} - z^{-3} = (z^3 - z^{-3}) - 3(z - z^{-1})$, and since $(z^3 - z^{-3}) = i2\sin 3\theta$, the required result follows.
2. By de Moivre, $(\cos \pi/4 + i\sin \pi/4)^2 = \cos \pi/2 + i\sin \pi/2 = i$. Hence $\sqrt{i} = \cos \pi/4 + i\sin \pi/4 = (1+i)/\sqrt{2}$.

EXERCISES 5D

1. $$\frac{4+i7}{3-i} = \frac{4+i7}{3-i}\,\frac{3+i}{3+i} = \frac{5+i25}{10} = \frac{1+i5}{2}.$$

The modulus is $\sqrt{\dfrac{1+i5}{2}\,\dfrac{1-i5}{2}} = \sqrt{\dfrac{1+25}{4}} = \sqrt{13/2}.$

2. $$\frac{(\cos 3\theta + i\sin 3\theta)^2}{(\cos\theta - i\sin\theta)^5} = \frac{(z^3)^2}{(z^{-1})^5} = \frac{z^6}{z^{-5}} = z^{11} \qquad (\text{where } z = \cos\theta + i\sin\theta)$$
$$= \cos 11\theta + i\sin 11\theta, \qquad \text{from De Moivre's Theorem.}$$

EXERCISES 6A

1. P is $(5, 4)$ and Q is $(3, 2)$ so the gradient of PQ is $(4-2)/(5-3) = 1$. Line of gradient 1 passing throught the origin is $y = x$.
2. Let the line be $y = mx + c$, intersecting the y-axis at $y = c$. But $m = \tan \pi/4 = 1$ and the line passes through $(2, 8)$ so $8 = 2 + c$. Hence $c = 6$.

EXERCISES 7A

1. (a) $x=6$ or 4; (b) $x=4/5$ or $3/7$; (c) $x=2+i$ or $2-i$.
2. The discriminant is $72^2-4.16\cdot C$ and roots are equal if the discriminant is zero, so $C=81$.

EXERCISES 7B

1. The graph similar to curve 1 of Fig. 7.5, cuts the x-axis ($y=0$) at (approximately) $x=1.78$, 1.22, -0.22, -0.86, and -2.14.
2. Plot $y=\tan 2x$ and (the straight line) $y=2x+2$; the intersection of the straight line with the tangent curves give values of x which satisfy the equation. But the tangent curve $\to \pm\infty$ whenever $x\to(n\pi\pm\pi/4)$, so that the line intersects it infinitely many times; between $x=0$ and $x=\pi$ it intersects at (approximately) $x=0.637$ and 2.281 radians.

EXERCISES 7C

1. $x>2$.
2. If $-2<x^2+3x$ then $x^2+3x+2>0$, i.e. $(x+1)(x+2)>0$ so both factors are +ve or both are −ve and $x<-2$ or $x>-1$. If $x^2+3x<10$ then $x^2+3x-10<0$, i.e. $(x+5)(x-2)<0$ so the factors are of opposite sign and $-5<x<2$. The two conditions together give $-5<x<-2$ or $-1<x<2$.

EXERCISES 8A

1. The vector $(-9, 12)$ has direction ϕ where $\tan\phi=-12/9=-4/3$. Let the required vector be (a, b); then $\sqrt{a^2+b^2}=10$ and $b/a=3/4$. Hence the required vector is $(8, 6)$ (or $(-8, -6)$).
2. The mid-point of A and B is $((3-1)/2, (7-3)/2)=(1, 2)$. The mid-point of A and C is $((13+3)/2, (45+7)/2)=(8, 26)$. The vector joining the mid-points is $(8-1, 26-2)$, i.e. $(7, 24)$ which has magnitude $\sqrt{7^2+24^2}$, i.e. magnitude 25 and direction θ where $\tan\theta=24/7$. The vector joining C and B is $(13-(-1), 45-(-3))=(14, 48)$ which is twice the magnitude of the other but in the same direction.

EXERCISES 8B

1. If $\mathbf{D}$ is (x, y, z) then $\mathbf{B}+\mathbf{C}+\mathbf{D}=((-3+1+x), (-2-1+y), (-1+1+z))$. Hence $\mathbf{B}+\mathbf{C}+\mathbf{D}=\mathbf{A}=(1, 2, 3)$ if $-2+x=1$ so that $x=3$, and $-3+y=2$ so that $y=5$, and $z=3$. Hence $\mathbf{D}$ is $(3, 5, 3)$.
2. Let $\mathbf{R}$ be (p, q) and let $\mathbf{S}=\mathbf{R}+\mathbf{P}=(1+p, 7+q)$ where $|\mathbf{S}|=64$ so that

$(1+p)^2+(7+q)^2=64^2=4096$. The direction of $\mathbf{Q}$ is given by $\tan^{-1}(5/5)$, i.e. by $\tan^{-1}1$, so the direction of $\mathbf{S}$ is given by $\tan^{-1}(-1)$ and $(7+q)/(1+p)=-1$. Thus $(1+p)^2=(7+q)^2=2048$, which leads to $\mathbf{R}=(44.25, 38.25)$.

EXERCISES 8C

1. $|\mathbf{A}|=\sqrt{4+4+1}=3$; $|\mathbf{B}|=\sqrt{36+0+9}=3\sqrt{5}$;

   The scalar product is $2\cdot(-6)+2\cdot(0)+(-1)\cdot 3=-15$;
   The vector product is $(2\cdot 3-0)\hat{x}+(3\cdot 2-6\cdot 1)\hat{y}+(2\cdot 0+2\cdot 6)\hat{z}$, i.e., $(6, 0, 14)$;
   The angle is

$$\cos^{-1}\frac{\mathbf{A}\cdot\mathbf{B}}{|\mathbf{A}|\cdot|\mathbf{B}|}=\frac{-15}{3\cdot 3\sqrt{5}}=-\frac{\sqrt{5}}{3} \quad \text{between 0 and } \pi,$$
$$=2.412 \text{ radians } (=138.19°).$$

2. The scalar product is $3\cdot 2+8\cdot(-1)+(-2)\cdot(-1)=0$. Hence the angle between the vectors is $\cos^{-1}0=\pi/2$.

EXERCISES 8D

1. $$\mathbf{B}=\mathbf{R}\cdot\mathbf{A}=\begin{pmatrix}1 & -1\\ 1 & 1\end{pmatrix}\begin{pmatrix}a\\ b\end{pmatrix}=\begin{pmatrix}a-b\\ a+b\end{pmatrix}.$$

$$\mathbf{C}=\mathbf{R}\cdot\mathbf{B}=\begin{pmatrix}1 & -1\\ 1 & 1\end{pmatrix}\begin{pmatrix}a-b\\ a+b\end{pmatrix}=\begin{pmatrix}(a-b)-(a+b)\\ (a-b)+(a+b)\end{pmatrix}=\begin{pmatrix}-2b\\ 2a\end{pmatrix}.$$

   The direction of $\mathbf{A}$ is $\tan^{-1}b/a$; the magnitude of $\mathbf{A}$ is $\sqrt{a^2+b^2}$.
   The direction of $\mathbf{C}$ is $\tan^{-1}(-2a/2b)=\tan^{-1}(-a/b)$; the magnitude of $\mathbf{C}$ is

$$\sqrt{4b^2+4a^2}=2\sqrt{a^2+b^2}.$$

   Hence $\mathbf{C}$ is perpendicular to A and twice its size, i.e. two applications of $\mathbf{R}$ rotate a vector through $\pi/2$ and double it. Thus, one application of $\mathbf{R}$ probably rotates by $\pi/4$ and multiplies by $\sqrt{2}$.

   Test: if $\mathbf{A}$ is $\binom{1}{1}$ its magnitude is $\sqrt{2}$ and its direction is $\theta_A=\tan^{-1}1=\pi/4$; then $\mathbf{B}$ is $\binom{0}{2}$, its magnitude is 2, and its direction along the $y-$axis has $\theta_B=\pi/2$; and $\mathbf{C}$ is $\binom{-2}{2}$ with the magnitude $2\sqrt{2}$ and direction $\theta_C=\tan^{-1}(-1)=3\pi/4$.
2. If $\mathbf{P}=\begin{pmatrix}1 & 0\\ 2 & -1\end{pmatrix}$ then $\mathbf{P}^2=\begin{pmatrix}1 & 0\\ 0 & 1\end{pmatrix}$; if $\mathbf{Q}=\begin{pmatrix}2 & -2\\ 1 & -2\end{pmatrix}$ then $\mathbf{Q}^2=\begin{pmatrix}2 & 0\\ 0 & 2\end{pmatrix}$ so that $\mathbf{P}^2+\mathbf{Q}^2=\begin{pmatrix}3 & 0\\ 0 & 3\end{pmatrix}$; But $\mathbf{P}+\mathbf{Q}=\begin{pmatrix}3 & -2\\ 3 & -3\end{pmatrix}$ so that $(\mathbf{P}+\mathbf{Q})^2=\begin{pmatrix}3 & 0\\ 0 & 3\end{pmatrix}$ QED.

EXERCISE 8E

(a) 7; (b) -4; (c) $\begin{vmatrix} 2 & 7 & 3 \\ 1 & 3 & 5 \\ 4 & 6 & -6 \end{vmatrix} = 2\begin{vmatrix} 3 & 5 \\ 6 & -6 \end{vmatrix} - 7\begin{vmatrix} 1 & 5 \\ 4 & -6 \end{vmatrix} + 3\begin{vmatrix} 1 & 3 \\ 4 & 6 \end{vmatrix} = 68.$

EXERCISES 8F

1. $$\begin{vmatrix} 7 & 6 & -12 \\ 14 & -8 & 3 \\ -7 & 2 & -9 \end{vmatrix} = \begin{vmatrix} 7 & 6 & -12 \\ 0 & -4 & -15 \\ -7 & 2 & -9 \end{vmatrix} = \begin{vmatrix} 7 & 6 & -12 \\ 0 & -4 & -15 \\ 0 & 8 & -21 \end{vmatrix}$$

$$= 7\begin{vmatrix} -4 & -15 \\ 8 & -21 \end{vmatrix} = 1428.$$

2. $$\mathbf{D} = \begin{vmatrix} 1 & 1 & 1 & x \\ 1 & 1 & x & 1 \\ 1 & x & 1 & 1 \\ x & 1 & 1 & 1 \end{vmatrix} = \begin{vmatrix} 1 & 1 & 1 & x-1 \\ 1 & 1 & x & 1-x \\ 1 & x & 1 & 0 \\ x & 1 & 1 & 0 \end{vmatrix} = \begin{vmatrix} 1 & 1 & 0 & x-1 \\ 1 & 1 & x-1 & 1-x \\ 1 & x & 1-x & 0 \\ x & 1 & 0 & 0 \end{vmatrix}$$

$$= \begin{vmatrix} 1 & 0 & 0 & x-1 \\ 1 & 0 & x-1 & 1-x \\ 1 & x-1 & 1-x & 0 \\ x & 1-x & 0 & 0 \end{vmatrix}.$$

Take the factor $(x-1)$ out of columns 1, 2, and 3.

$$\mathbf{D} = (x-1)^3 \begin{vmatrix} 1 & 0 & 0 & 1 \\ 1 & 0 & 1 & -1 \\ 1 & 1 & -1 & 0 \\ x & -1 & 0 & 0 \end{vmatrix};$$

then $C_4 - C_1$ gives

$$\mathbf{D} = (x-1)^3 \begin{vmatrix} 1 & 0 & 0 & 0 \\ 1 & 0 & 1 & -2 \\ 1 & 1 & -1 & -1 \\ x & -1 & 0 & -x \end{vmatrix}.$$

Hence

$$\mathbf{D}=(x-1)^3\begin{vmatrix} 0 & 1 & -2 \\ 1 & -1 & -1 \\ -1 & 0 & -x \end{vmatrix}=(x-1)^3\begin{vmatrix} 0 & 1 & -2 \\ 0 & -1 & -x-1 \\ -1 & 0 & -x \end{vmatrix}$$

$$=(x-1)^3(x+3).$$

EXERCISE 9A

(a) $x=11$, $y=2$; (b) $x=1$, $y=0$; (c) insoluble (parallel lines).

EXERCISES 9B

1. $\Delta=12$, $\Delta_1=-24$, $\Delta_2=12$, $\Delta_3=6$, so $x=-2$, $y=1$ and $z=1/2$.
2. $\Delta=-33$, $\Delta_1=-33$, $\Delta_2=66$, $\Delta_3=99$, $\Delta_4=132$, so $w=1$, $x=-2$, $y=3$ and $z=-4$.

EXERCISE 9C

The equation is $y=2-x-2x^2+x^3$, and the graph is similar to curve 1 of Fig. 7.4. The local maximum is at $(-0.22, 2.30)$ and the local minimum at $(1.55, -0.63)$.

EXERCISE 9D

For the matrix

$$\mathbf{M}=\begin{pmatrix} 2 & 3 & 4 \\ 1 & -1 & 8 \\ 3 & 1 & 12 \end{pmatrix},$$

the transpose matrix

$$\mathbf{M}'=\begin{pmatrix} 2 & 1 & 3 \\ 3 & -1 & 1 \\ 4 & 8 & 12 \end{pmatrix},$$

the adjoint matrix,

$$\text{adj }\mathbf{M}=\begin{pmatrix} -20 & -32 & 28 \\ 12 & 12 & -12 \\ 4 & 7 & -8 \end{pmatrix},$$

and, as in exercises 9B, number 1, the determinant $|\mathbf{M}|=12$ giving $x=-2$, $y=1$, and $z=1/2$.

EXERCISES 10A

1. (a) $18x^2+8x+2$; (b) $2x-2x^{-3}$; (c) $\pi x\sec^2(\pi x^2/2)$;
   (d) $3(1+3x)(1+2x+3x^2)^{1/2}$.
2. (a) $-e^{-x}(\cos x^2+2x\sin x^2)$; (b) $-2x(1+x^2)^{-2}$; (c) $-(1-x^2)^{-\frac{1}{2}}(1+x)^{-1}$;
   (d) $\ln a\cdot a^{x\sin x}(\sin x+x\cos x)$.
3. $\mathrm{d}B/\mathrm{d}W=(1.78)(0.73)W^{-0.27}$ and $\mathrm{d}B/\mathrm{d}W=2$ when $W=0.2022$ kg.

EXERCISES 10B

1. (a) $\mathrm{d}y/\mathrm{d}x=0$ when $x=0.863$ giving a minimum ($\mathrm{d}^2y/\mathrm{d}x^2$ +ve) or $x=-0.463$ giving a maximum ($\mathrm{d}^2y/\mathrm{d}x^2$ −ve).
   (b) $\mathrm{d}y/\mathrm{d}x=0$ when $x=0.073$ (minimum), 0 (maximum), or -3.448 (minimum).
2. (a) If rectangle has sides a and b then $2a+2b=40$ so $b=20-a$ and area $A=ab=20a-a^2$. Then $\mathrm{d}A/\mathrm{d}a=0$ when $a=10$, a maximum, giving $A=100\,\mathrm{m}^2$.
   (b) The circle of radius r has circumference $2\pi r=40$, area $=127.3\,\mathrm{m}^2$.
   (c) The perimeter is $a+2b=40$ and area $A=20a-a^2/2$ giving $\mathrm{d}A/\mathrm{d}a=0$ when $a=20$ (and $b=10$) a maximum, with $A=200\,\mathrm{m}^2$.
3. For amplification A,

$$\frac{\mathrm{d}A}{\mathrm{d}x}=k\frac{a^2-x^2}{(a^2+x^2)^2} \quad \text{and} \quad \frac{\mathrm{d}^2A}{\mathrm{d}x^2}=-\frac{2kx(3a^2-x^2)}{(a^2+x^2)^3}.$$

   $\mathrm{d}A/\mathrm{d}x=0$ when $x=\pm a$. A is at a maximum when $x=a$. Mathematically, A is at a minimum when $x=-a$, but for real transistors in the real world the amplification must always be positive ($A\to 0$ as $x\to 0$ and as $x\to\infty$.)

EXERCISE 10C

$$\frac{\partial z}{\partial x}=\frac{2xy^4}{(x^2+y^2)^2}, \qquad \frac{\partial^2 z}{\partial x^2}=\frac{2y^6-6x^2y^4}{(x^2+y^2)^3},$$

$$\frac{\partial z}{\partial y}=\frac{2x^4y}{(x^2+y^2)^2}, \qquad \frac{\partial^2 z}{\partial y^2}=\frac{2x^6-6x^4y^2}{(x^2+y^2)^3};$$

$$\frac{\partial^2 z}{\partial x\partial y}=\frac{\partial^2 z}{\partial y\partial x}=\frac{8x^3y^3}{(x^2+y^2)^3}.$$

EXERCISES 10D

1. (a) $x^{10}+x^7-x^4+C$; (b) $\dfrac{x^4+x^2}{2}+C$; (c) $\dfrac{1}{\alpha}\sec \alpha x+C$; (d) $C-\frac{3}{4}\cos\dfrac{2x}{3}$;
   (e) $\ln(1+x)+C$; (f) $\tan^{-1}x+C$.

2. $\displaystyle\int_0^1 (4x^4-x+7)\,dx=\left[\frac{4x^5}{5}-\frac{x^2}{2}+7x\right]_0^1=(\tfrac{4}{5}-\tfrac{1}{2}+7)=7.3.$

EXERCISES 10E

1. (a) Substitute $z=7+5\cos x$, $dz/dx=-5\sin x$, so

$$I_1=C-\tfrac{1}{5}\ln(7+5\cos x).$$

   (b) Substitute $z=x^3+x^2+1$, etc. so

$$I_2=C-\frac{1}{2(x^3+x^2+1)^2}.$$

   (c) Substitute $z=x^3$ so that $d/dx(\sin z)=\cos z\,dz/dx$ and $I_3=C+\frac{1}{3}\sin x^3$.

2. (a) Integrate by parts to give

$$I_1=\frac{(x+1)^2}{2}\ln(x+1)+\frac{(x+1)^2}{4}+C.$$

   (b) Integrate by parts to give

$$I_2=\frac{x^3(x^3+1)^6}{18}-\frac{(x^3+1)^7}{126}+C.$$

3. (a) $I_1=\displaystyle\int\frac{dx}{\sqrt{(x+2)^2+1}}$. Substitute $\tan\theta=x+2$, leading to

$$I_1=\int\sec\theta\,d\theta=\ln(\sec\theta+\tan\theta)+C=\ln(\sqrt{x^2+4x+5}+x+2)+C.$$

   (b) $\dfrac{x}{x^2+4x+5}=\dfrac{1}{2}\dfrac{2x+4}{x^2+4x+5}-\dfrac{2}{(x+2)^2+1}$

   so

$$I_2=\tfrac{1}{2}\ln(x^2+4x+5)-2\tan^{-1}(x+2)+C.$$

(c) $$\frac{x}{\sqrt{x^2+4x+5}}=\frac{1}{2}\frac{2x+4}{\sqrt{x^2+4x+5}}-\frac{2}{\sqrt{x^2+4x+5}}$$

so

$$I_3=(x^2+4x+5)^{\frac{1}{2}}-2\ln(\sqrt{x^2+4x+5+x+2})+C.$$

4. Curves intersect at (0, 0) and when $y^4=16\cdot\frac{1}{2}y$, i.e. when $y=2$. Thus points of intersection are (0, 0) and (1, 2). A graph shows that the area of overlap is

$$\int_0^1 2x^{\frac{1}{2}}\,dx-\int_0^1 2x^2\,dx=\tfrac{2}{3}.$$

5. Integrating by parts,

$$I=\int e^{3\theta}\cos 5\theta\,d\theta=\tfrac{1}{5}e^{3\theta}\sin 5\theta-\tfrac{3}{5}\int e^{3\theta}\sin 5\theta\,d\theta$$

$$=\tfrac{1}{5}e^{3\theta}\sin 5\theta+\tfrac{3}{25}e^{3\theta}\cos 5\theta-\tfrac{9}{25}\int e^{3\theta}\cos 5\theta\,d\theta.$$

Hence

$$\tfrac{34}{25}I=\tfrac{1}{5}e^{3\theta}\sin 5\theta+\tfrac{3}{25}e^{3\theta}\cos 5\theta$$

and

$$I=\tfrac{5}{34}e^{3\theta}\sin 5\theta+\tfrac{3}{34}e^{3\theta}\cos 5\theta+C.$$

EXERCISES 11A

1. (a) $a=2$, $d=1$, the tenth term is $6\frac{1}{2}$, and the sum to ten terms is $42\frac{1}{2}$;
   (b) $a=53$, $d=-3$, the tenth term is 26, and the sum to ten terms is 395.
2. $d=-5$.

EXERCISES 11B

1. (a) $a=1.000$, $r=0.9$, $a_6=0.590$, $S_6=4.686$, $S_\infty=10.000$;
   (b) $a=1.000$, $r=-1.100$, $a_6=-1.611$, $S_6=-0.367$; no sum to infinity;
   (c) $a=81$, $r=\frac{1}{3}$, $a_6=0.333$, $S_6=121.333$, $S_\infty=121.500$.
2. $a=1$, $r=2x$, $S=a/(1-2x)$ provided that $|2x|<1$, i.e. $-0.5<x<0.5$.

EXERCISE 11C

(a) $u_n=n(n+1)$, $v_n=n(n+1)(n+2)$, $\Sigma u_n=\frac{1}{3}(v_n-v_0)$ and $v_0=0$, so $S_{10}=440$.
(b) $u_n=1/n(n+1)$, $v_n=1/n$, $\Sigma u_n=v_1-v_{n+1}=1-1/(n+1)$, so $S_{10}=10/11$.

EXERCISES 11D

1. $\sin^2 2x = \frac{1}{2}(1-\cos 4x) = 4x^2 - \frac{16}{3}x^4 + \frac{128}{45}x^6 - \frac{256}{315}x^8 + \cdots$
2. $\sin(\pi/6) = 0.50000$ and $\cos(\pi/6) = 0.86603$. Expanding to three terms,

$$\sin(\pi/6) = (\pi/6) - (\tfrac{1}{6})(\pi/6)^3 + (\tfrac{1}{120})(\pi/6)^5 = 0.50000;$$
$$\cos(\pi/6) = 1 - (\tfrac{1}{2})(\pi/6)^2 + (\tfrac{1}{24})(\pi/6)^4 = 0.86605.$$

EXERCISES 11E

1. (a) $\cosh^2 A + \sinh^2 A = \frac{1}{4}((e^A + e^{-A})^2 + \frac{1}{4}(e^A - e^{-A}))^2$

$$= \tfrac{1}{4}((e^{2A} + 2 + e^{-2A}) + (e^{2A} - 2 + e^{-2A})) = \cosh 2A.$$

(b) $\tanh^2 A = \dfrac{\sinh^2 A}{\cosh^2 A} = \dfrac{\cosh^2 A - 1}{\cosh^2 A} = 1 - \operatorname{sech}^2 A.$

2. (a) $\dfrac{d}{dx}[\frac{1}{2}(e^x - e^{-x})] = \frac{1}{2}(e^x + e^{-x}) = \cosh x;$

(b) $\dfrac{d}{dx}[\frac{1}{2}(e^x + e^{-x})] = \frac{1}{2}(e^x - e^{-x}) = \sinh x;$

(c) $\dfrac{d}{dx}\left(\dfrac{\sinh x}{\cosh x}\right) = \dfrac{\cosh^2 x - \sinh^2 x}{\cosh^2 x} = \dfrac{1}{\cosh^2 x} = \operatorname{sech}^2 x.$

EXERCISE 12A

(a) Separating variables,

$$\int y\,dy = \int \frac{1+x}{x}\,dx$$

so $y^2/2 = \ln x + x + C'$, so $y = \sqrt{\ln x^2 + 2x + C}$.

(b)
$$\sin 2y \frac{dy}{dx} = [\cos(x+y) + \cos(x-y)]\sin 2x,$$

$$\therefore\ 2\sin y \cos y \frac{dy}{dx} = 2\cos x \cos y \cdot 2\sin x \cos x = 4\cos y \sin x \cos^2 x.$$

$\therefore$ separating variables,

$$\int \sin y\,dy = \int 2\sin x \cos^2 x\,dx$$

which gives $\cos y = \frac{2}{3}\cos^3 x + C$ or $y = \cos^{-1}(\frac{2}{3}\cos^3 x + C)$.

EXERCISES 12B

1. (a) Write the equation as

$$x^2\frac{dy}{dx}-y=x^3\,e^{x-1/x}.$$

Divide through by x^2 so the integrating factor is $e^{\int -1/x^2\,dx}=e^{1/x}$. Multiply through by $e^{1/x}$. Hence

$$\frac{d}{dx}(y\,e^{1/x})=x\,e^x.$$

This leads to

$$y=(x-1)\,e^{(x^2-1)/x}+C\,e^{-1/x}.$$

(b) The integrating factor is $e^{\int \cot\theta\,d\theta}=e^{\ln\sin\theta}=\sin\theta$. Multiply through by $\sin\theta$ and the equation yields $x=\operatorname{cosec}\theta(\ln\sin\theta+C)$.

(c) Integrating factor is $e^{-(\cos 2\theta)/2}$ and $x=\frac{1}{2}\theta^2\,e^{\cos^2\theta}+C\,e^{\cos^2\theta}$.

2. The mouse population has a breeding rate

$$\left(\frac{dm}{dt}\right)^+=Bm.$$

If there are H households then there are H cats which each catches k mice. Hence Hk mice are caught by cats and T are trapped so the rate of loss of mice is

$$\left(\frac{dm}{dt}\right)^-=-(Hk+T).$$

Thus the initial differential equation is

$$\frac{dm}{dt}=Bm-Hk-T.$$

(a) While the mouse population is stable, $dm/dt=0$ with $m=M_0$, say. Thus, $Hk=BM_0-T$ or $k=(BM_0-T)/H$ is the number of mice caught per cat.

(b) On doubling the cat population, the number of mice caught rises to $2Hk=2BM_0-2T$ and the differential equation becomes

$$\frac{dm}{dt}=Bm-2Hk-T=Bm-2BM_0+T.$$

The integrating factor is e^{-Bt} and the solution is

$$m=M_0-(M_0-T/B)(e^{Bt}-1).$$

m falls to zero after a time

$$\left(\frac{1}{B}\right)\ln\left(\frac{2BM_0-T}{BM_0-T}\right).$$

EXERCISES 12C

1. (a) $y=A\,e^{-3x}+B\,e^{-x}$.
   (b) $y=A\,e^{-4x}+B\,e^{-3x}$; but $y=5$ when $x=0$ so $A+B=5$, and $y=1$ when $x=1$ so $A\,e^{-4}+B\,e^{-3}=1$. Solving for A and B gives $y=28.84\,e^{-3x}-23.84\,e^{-4x}$.
2. The auxiliary equation is $D^2+3D+2=0$ so the complementary function is $A\,e^{-2x}+B\,e^{-x}$. The particular integral is $e^{3x}/(3^2+3\cdot3+2)=e^{3x}/20$, so the complete solution is $y=A\,e^{-2x}+B\,e^{-x}+e^{3x}/20$.
3. The complementary function is $y=A'\,e^{-1+ix/2}+B'e^{-1-ix/2}$. The particular integral comes from $\sin x/(4D^2+8D+5)$. Replace D^2 by $-(1^2)$, multiply by $(8D-1)/(8D-1)$, and again replace D^2 by -1. The particular integral is

$$\frac{(8D-1)\sin x}{-65}=\frac{\sin x-8\cos x}{65}$$

and the complete solution after rewriting the complementary function is

$$y=e^{-x}\left(\cos\left(\frac{x}{2}\right)+B\sin\left(\frac{x}{2}\right)\right)+(\sin x-8\cos x)/65.$$

EXERCISES 13A

1. 2.38. The erroneous s.d. value is

$$\sqrt{\frac{\Sigma(x_i-m)^2}{n}}=2.20.$$

2. 6069.2.

EXERCISE 13B

The first quartile (tenth place from top) is $q_1=100$ W; the third quartile (tenth place from bottom) is $q_3=40$ W. (NB: the median, q_2, is 60 W, so skewness is revealed by the difference between $q_1-q_2=40$ W and $q_2-q_3=20$ W; cf. exercises 13A, number 2 and Figs 13.1 and 13.4.)

EXERCISES 13C

1. It is stated that the distribution is normal (Gaussian) so use equation (13.10) with $\mu=50$ and $\sigma=20$. Tables such as those in section 14.12 give the cumulative probability of a result less than a given value of x. With $x=30$, equivalent to $\xi=1$ in equation (13.6), the tabulated value gives a probability of 0.159 of observing not more than 30% of flies being killed.
2. A Poisson distribution with mean $\mu=5$ (out of 100) gives (by direct calculation or, better, from tabulated values) a probability of 0.133 of making a count of 8 or more.
3. Assume a binomial distribution with a probability $p=0.1$ of success (for the customs officers) and $q=0.9$ of failure. Then in five trials the probability is 0.082 of two or more successes.

EXERCISE 13D

Analysis of the data shows (Table A.1) that, for n_i litters of size x_i, the total number of litters is $\sum n_i=60$ and the total of piglets is $\sum x_i n_i=360$. Hence the average litter size is 6.0 with standard deviation $s=1.636$. Any probability distribution $\phi(x)$ will give the fraction of live births to have been included in a litter of size x. Hence it is appropriate to compare the recorded figures of $n_i x_i$ with $360\,\phi(x)$ for various distributions. The situation is not one of two states so as binomial distribution is inappropriate. A Poisson distribu-

Table A.1 Piglet litters

x_i	n_i	$n_i x_i$	$(nx)_P$	$(nx)_G$	Number of females, f	$f/(n_i x_i)$
1	0	0	5.35	0.82	0	–
2	1	2	16.06	4.42	1	0.500
3	3	9	32.12	16.34	6	0.667
4	6	24	48.18	41.58	15	0.625
5	11	55	57.82	72.82	30	0.545
6	19	114	57.82	87.79	68	0.596
7	9	63	49.56	72.82	37	0.587
8	7	56	37.17	41.58	33	0.589
9	3	27	24.78	16.34	15	0.556
10	1	10	14.86	4.42	5	0.500
11	0	0	8.11	0.82	0	–
Totals:	60	360	351.84	359.75	210	

tion is given by

$$(nx)_P = 360\frac{e^{-6}6^x}{x!};$$

a Gaussian distribution is given by

$$(nx)_G = \frac{360}{1.636\sqrt{2\pi}}e^{-\frac{(x-6)^2}{2(1.636)^2}}.$$

The results are given in Table A.1. While the observed numbers are clearly not consistent with a Poisson distribution, they agree tolerably with the expected values for a Gaussian distribution, although they are somewhat too peaked. The results show that the sample is rather small for reliable conclusions to be drawn. This will be even more the case for attempted analysis of the distribution of female piglets among the litter sizes. Apart from this consideration, the distribution of females could only properly be compared to the distribution of the mixed litters if the sex ratio is independent of litter size. However, the last column of Table A.1 suggests that the fraction of females may fall as litter size increases; and this matter needs to be tested by correlation analysis as discussed in the next chapter.

EXERCISE 14B

(height in centimetres) = 5.80 (age in years) + 81.56.

EXERCISES 14C

$S_Q = 0.0289$, i.e. 0.87% of S_L.

The quadratic regression curve is $H = 72.51 + 8.34A - 0.14A^2$. This predicts heights of 110.7, 141.9, 166.1, 183.3, 193.5, 196.7 and 192.9 cm at ages 5, 10, 15, 20, 25, 30 and 35 years. Comparison with the values in Table 14.1 shows that the regression curve can be used to *interpolate* to plausible (though not very accurate) values; however, the *extrapolated* values are quite implausible and indeed the curve requires that the people shrink after the age of 30. Regression lines should be used for extrapolation only with great caution.

EXERCISE 14D

If the die is unbiased, the expected frequency for each number is 16.67. Calculation gives $\chi^2 = 10.398$. Thus the die just scrapes by the usual criterion of acceptability, but if money is involved it might be safest to obtain another die.

EXERCISE 14E

There are $5+1-2=4$ degrees of freedom and taking the null hypothesis 'H_0: the measurements are not incompatible (i.e. the measurements come from the same population)' so that $\mu_1=\mu_2$ in equation (14.20), calculation gives $t_4=1.11$. Using a 'one-tail' test, tables of the t-distribution show that there is a chance greater than 10% of making a single measurement which exceeds the mean by as much as that reported. Thus H_0 can be upheld.

EXERCISE 14F

The individual means of the four groups are 2.30, 2.20, 2.60 and 2.20 and the grand mean is $X=2.30$. The null hypothesis and variance calculations are similar to those of example 14(g) and details are given in Table A.2. The result is $F_{3,21}=8.734$ which is much larger than the 5% probability value of $F_{3,21}$ (cf. 4. of the answer to example 14(g)). This indicates a chance of only about 1 in 2000 of obtaining such discrepant results.) Therefore the hypothesis that all the samples are from the same population (i.e. are of the same purity) must be rejected. (NB: within each preparation, results are very reproducible, but preparation 3 has a mean distinctly different from the means of the other preparations.)

Table A.2 Analysis of variance: dehydrogenase rate constants

Source	Degrees of freedom (f)	Sum of squares (SSq)	Variance	Variance ratio (F)
		$\sum n_j(x_j-X)^2$	$s^2=\dfrac{SSq}{f}$	$\dfrac{s_1^2}{s_2^2}$
Between groups	$f_1=4-1=3$	0.6000	$\dfrac{0.600}{3}=0.200$	$\dfrac{0.20}{0.0229}=8.734$
Within groups	$f_2=4+6+4+7$ $=21$	0.4799	$\dfrac{0.48}{21}=0.0229$	

EXERCISES 15B

1. The function is even so $b_n=0$ for all n.

$$f(x)=1-\frac{2}{3\pi}+\sum\frac{4}{\pi}\left(\frac{\cos n}{n^2}-\frac{\sin n}{n^3}\right)\cos nx.$$

2. $$f(x)=\frac{\sinh \pi}{\pi}\left(1+\sum(-1)^n 2(n^2+1)^{-1}\ (\cos nx-n\sin nx)\right).$$

3. Write the function as

$$P'(t)=\begin{cases} a+b\cos^2\left(\dfrac{3t}{2}\right) & \text{for } -\dfrac{\pi}{3}\leqslant t\leqslant\dfrac{\pi}{3}; \\ a & \text{for } -\pi<t\leqslant-\dfrac{\pi}{3} \text{ and for } \dfrac{\pi}{3}<t\leqslant\pi. \end{cases}$$

The function is now even in the range $-\pi<t<\pi$ and in its Fourier series expansion, $f(t)$, $b_n=0$ for all n. Note that

$$2\cos^2\left(\frac{3t}{2}\right)=\cos 3t+1.$$

Then

$$a_0=a+b/6 \qquad a_1=7\sqrt{3}b/8\pi \qquad a_2=\sqrt{3}b/4\pi+3b/5\pi$$
$$a_3=b/6 \qquad a_4=3b/7\pi-\sqrt{3}b/8\pi.$$

EXERCISE 15C

(a) Taking a_0 and a_n from the answer to exercises 15B, number 1, $g_0=a_0$ and since the function is even $g_n=g_{-n}=a_n/2$.

$$g_0=1-2/3\pi;$$
$$g_n=(2/\pi n^2)(\cos n-(\sin n)/n)=g_{-n}$$

since $\sin(-n)=-\sin n$.

(b) Taking a_0, a_n and b_n from the answer to exercises 15B, number 2,

$$g_0=a_0=(\sinh\pi)/\pi.$$
$$g_n=\tfrac{1}{2}(a_n-ib_n)=(-1)^n\sinh\pi(\pi(n^2+1))^{-1}(1+in)$$
$$g_{-n}=\tfrac{1}{2}(a_n+ib_n)=(-1)^n\sinh\pi(\pi(n^2+1))^{-1}(1-in)$$

EXERCISE 15D

The function is even, so $b_n=0$ for all n. Put $\alpha_n=\pi np/Q$. Then

$$f(t)=2p^2\left[(1/4Q)+\sum(\alpha_n^{-1}\sin\alpha_n+\alpha_n^{-2}\cos\alpha_n)\cos(\alpha_n t/p)\right].$$

EXERCISE 15E

$F(\omega)=2(1-\cos\omega p)/\omega^2.$

I'm very well acquainted (now) with matters mathematical,
I understand equations, both simple and quadratical,
About binomial Theorem I'm teeming with a lot of news,
With many cheerful facts about the square of the hypotenuse.
I'm very good at integral and differential calculus;
I know the scientific names of beings animalculous.
In short, in matters vegetable, animal, and mineral,
I am the very model of a modern Major-General.

W. S. Gilbert, *The Pirates of Penzance*

Index